2 I 23

Lending Library

Camden
LEISURE & COMMUNITY

This book is due for return on or before the date stamped below. If not required by another reader it may be renewed.

Queens Crescent Library
165 Queens Crescent
NW5 4HH
Tel: 020 7974 4444

Fines are charged on overdue books. Please bring your ticket with you.

08/02			
14 SEP 2002			
01 OCT 2003			
18 MAY 2005			

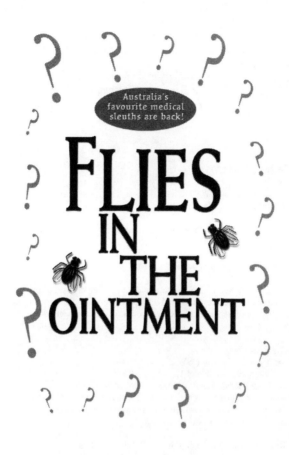

Australia's favourite medical sleuths are back!

FLIES IN THE OINTMENT

Medical Quacks, Quirks and Oddities

Dr Jim Leavesley & Dr George Biro

Illustrations by Adam Yazxhi

📚HarperCollins*Publishers*

*This book is dedicated
by Jim Leavesley to his stepchildren,
Jane and John Packham.
And by George Biro to the memory of his parents,
Renata and Paul Biro, and his stepfather, George Furer.*

HarperCollins*Publishers*

First published in Australia in 2001
by HarperCollins*Publishers* Pty Limited
ACN 009 913 517
A member of the HarperCollins*Publishers* (Australia) Pty Limited Group
http://www.harpercollins.com.au

HarperCollins*Publishers*
25 Ryde Road, Pymble, Sydney NSW 2073, Australia
31 View Road, Glenfield, Auckland 10, New Zealand
77–85 Fulham Palace Road, London W6 8JB, United Kingdom
Hazelton Lanes, 55 Avenue Road, Suite 2900, Toronto, Ontario M5R 3L2
and 1995 Markham Road, Scarborough, Ontario M1B 5M8, Canada
10 East 53rd Street, New York NY 10022, USA

National Library of Australia Cataloguing-in-Publication data:

Leavesley, J. H. (James H.), 1929- .
    Flies in the ointment : medical quacks, quirks and oddities.
    Bibliography.
    ISBN 0 7322 6933 4.
    1. Celebrities - Health and hygiene - Anecdotes. 2. Quacks
    and quackery - History - Anecdotes. 3. Medicine - History -
    Anecdotes. I. Biro, George. II. Title.
610.92

Every effort has been made by the authors to contact copyright
holders of quoted material contained herein. Any copyright holders
who may have been overlooked are welcome to contact the publishers.

Printed and bound in Australia by Griffin Press on 79gsm Bulky Paperback
9 8 7 6 5 4 3 2 1
04 03 02 01

ACKNOWLEDGMENTS

Jim Leavesley is grateful to the personnel of *Australian Doctor*, especially Nicole MacKee, for their encouragement and help, various broadcasters at the ABC in Perth and the Science Unit for their enthusiasm, especially Ted Bull, and the editorial staff of HarperCollins for their patience, especially Alison Urquhart.

George Biro thanks all those people who advised him in writing these stories, the staff of *Australian Doctor* who printed most of them, and Nicholas Guoth who helped choose them.

CONTENTS

Chapter 4

Chapter 5

Chapter 6

PREFACE

To help satisfy a natural curiosity that many people seem to have about "things medical", Jim Leavesley and George Biro have once again exposed their vanities, prejudices and morbid curiosity to bring together their third collection of essays on clinical oversights and underestimates, pathological vagaries and oddities, and medical bungles and botches. Some of the stories are well known, some more obscure, but all record a rather offbeat side of medicine — that aspect which appeals to our sense of fascinated disapproval.

As before, most of the pieces have appeared in the medical magazines *Australian Doctor* or *Medical Observer*, or, in Jim Leavesley's case, on ABC radio.

A difference this time is that Jim Leavesley has been allowed free range to his natural verbosity, so his pieces are longer but fewer than those of George Biro's more succinct and disciplined offerings. Rather surprisingly, they both passed the post together with about the same number of words.

Jim Leavesley was born and educated in the northern English seaside holiday resort of Blackpool. He graduated in medicine in Liverpool in 1953 and emigrated to Western Australia in 1957. After over 30 years in general practice he retired to Margaret River to take up the much more chancy occupation of writing about medical history, which for the past 10 years he has done on a fortnightly basis for the medical newspaper *Australian Doctor*. He is also a weekly presenter of the subject on ABC radio in WA, and an irregular guest on ABC Radio National's "Ockham's Razor". Before his partnership with George Biro, he had published five books on the subject.

George Biro was born in Budapest to an Italian mother and Hungarian father. The family migrated to Australia in 1947. He was nine when he arrived and could speak no English, but his application to study knew no closed season and he went on to graduate in medicine at Sydney University in 1963. George started his professional life as a GP/anaesthetist, but since 1990 he has been a freelance medical writer.

Together with *What Killed Jane Austen?* and *How Isaac Newton Lost His Marbles*, this is the third book George Biro has co-authored with Jim Leavesley.

Dr Jim Leavesley   *Dr George Biro*

Dead flies cause the ointment of the apothecary to send forth a
stinking savour ...

(ECCLESIASTES 10:I)

Romeo: I do remember an apothecary
And hereabouts he dwells — whom late I noted
In tatter'd weeds, with overwhelming brows,
Culling of simples; meagre were his looks.
Sharp misery had worn him to the bones

(ROMEO AND JULIET III:V)

See one physician like a sculler plies
The patient lingers and by inches dies;
But two physicians like a pair of oars
Waft him more swiftly to the Stygian shores

(ANON)

Chapter 1

PIONEERS

SIR NORMAN GREGG AND GERMAN MEASLES

In 1915, just after graduating with First Class Honours, Dr Norman Gregg enlisted in the Royal Army Medical Corps. Had it not been for the First World War, he might have played Davis Cup tennis for Australia. As it was, he won the Military Cross. In late 1940, with Australia at war, Dr Gregg was the senior eye doctor at Sydney's Royal Alexandra Hospital for Children.

In his own practice he was finding an unusual number of birth defects. Colleagues confirmed his impression, but no-one knew the cause. Then two of the mothers mentioned that they had had German measles (rubella) early in their pregnancies. Gregg asked colleagues around Australia to help him follow this lead. He gathered histories of 78 children born with cataracts (opacities in the lens of the eye affecting vision) in early 1941; 13 were his own patients. Many of the children had other serious defects as well.

Gregg suspected some sort of poison or infection. Most of the pregnancies had been normal, except for one thing: 68 of the 78 mothers had had rubella in the first or second month, or just before confirmation of the pregnancy. There had been an epidemic of rubella in 1940.

In October 1941, Gregg read his historic paper "Congenital cataract following German measles in the mother". By now 15 of the 78 babies (almost 20%) had already died. He noted that many also had serious heart defects and were of "small size, ill-nourished and difficult to feed". An editorial in the *Medical Journal of Australia* supported Gregg's findings: "The series is so striking and the sight of the children is so seriously affected that the facts must be made known ... "

With foresight, Gregg suggested "other defects are not evident now, but will show up as development proceeds". Two years later, reports from South Australia added deafness to the list.

For several years, medical authorities overseas did not accept Gregg's ideas, but in 1947 the *New England Journal of Medicine* ran an editorial on Gregg's findings. In 1951, the Australian statistician Professor Oliver Lancaster extended Gregg's work by tracing outbreaks of deafness in New South Wales back to the turn of the century. He blamed these on epidemics of rubella in pregnancy.

There was a worldwide outbreak of rubella in 1964–65, with 20 000 affected American infants. Till then, the distinction of rubella from other rashes had been purely clinical. Now came diagnostic tests showing that children infected before birth remained infected for months, even years. This explained the onset of cataracts and other defects after birth.

Gregg was knighted in 1953 and died in 1966 at the age of 74.

By 1970, Australia had a vaccine against rubella. From 1977, the number of children born deaf fell. By 1983, 96% of pregnant Australian women were immune to rubella, compared with 82% in 1971.

In 1989, Australia introduced vaccination of young children with the combined measles-mumps-rubella vaccine, while continuing vaccination of teenaged schoolgirls. Countries using widespread vaccination have reduced the rate of infection of the newborn and associated disabilities.

Sir Macfarlane Burnet assessed Gregg's work as "the most important contribution ever made to medicine in Australia". The *Australian Encyclopedia* describes it in even broader terms as "one of the milestones in the history of medicine".

(GB)

WILLIAM HARVEY AND THE CIRCULATION OF THE BLOOD

In 1628 a slim 72-page medical treatise was published in Frankfurt, Germany. At the time it was generally ignored by the profession, but over the years it has come to be regarded as arguably medical literature's greatest book. It was dedicated to Charles I of England and was less than snappily entitled *Excercitatio Anatomica De Motu Cordis et Sanguinis in Animalibus (The Anatomical Treatise on the Movement of the Heart and Blood in Animals)*, or De Motu Cordis for short. In it the author reached the then heretical conclusion that "the movement of the blood is constantly in a circle, and is brought about by the beat of the heart".

The author was William Harvey (1578–1657) and his description of the circulation of the blood allowed medicine to move out of the Dark Ages and medicinal practice to open up into the discipline we now know.

Before Harvey's time it was certainly known that blood moved, no doubt about that, but it was thought to do so in an ebbing and flowing manner, initiated by dilatation of the heart and blood vessels and then passing from the heart's right to left ventricle through minute pores in the septum which separated them. True enough, it was admitted these pores had not be seen, but it was felt they must be there nonetheless.

**The genius, Dr William Harvey, and his
"If you want something done check it a million times"
approach to experiments on the heart.**

Then along came the Belgium anatomist Andreas Vesalius (1514–1564), who arrived on the scene almost a hundred years before Harvey. Vesalius showed that these microscopic heart channels were a figment of the imagination. Predictably, contemporary medical thinking dismissed this opinion. The almost universal and well-entrenched view was that, besides the microscopic pores, the liver was the main organ in the blood system, a fiction which had been held dear, with sundry other medical myths, since the time of Galen, the Greek physician, 1400 years previously. It would take a courageous man to question such a lineage.

But came the moment, came the man, and William Harvey had the drive to be just such an adversary.

The genius of Harvey was that, unlike his predecessors, step by step he double-checked repeatable experiments before reaching the irrefutable conclusion that the heart was the crucial driving force of the blood system. Further, he demonstrated that the heart and veins had valves preventing backflow, and his experiments included noting the effect of ligation of the arteries and veins at various points. He compared the sequence of events within the hollow muscular organ, the heart, to a flintlock; the flint strikes the steel which ignites the powder to cause an explosion and thus ejects the bullet. He calculated the volume of blood at each heartbeat and showed its total amount would pass through the system in a comparatively short time.

Unfortunately, due to his terrible handwriting, *De Motu Cordis* contains a number of errors perpetuated by a baffled editor. Notes still preserved in the British Museum show a curious mixture of Latin and English held together by a lattice of lines, arrows and erasures. Transcription must have defied all but the most adept proofreader. Nonetheless, Harvey's publishers managed to get the basic stuff right and follow his dictum, "I avow myself the partisan of truth alone."

The singular result was that for the first time someone had demonstrated the anatomy of the circulatory system. But with a reticence typical of great researchers, Harvey hesitated to draw cast-iron conclusions, stating merely that whether the whole process was "for the sake of nourishment or for the communication of heat, is not certain".

What of the man himself? Harvey was born in Folkstone on 1 April 1578, the eldest of seven sons of Thomas Harvey, mayor of the town. Five of the sons became rich turkey merchants in the City of London. William alone pursued an acedemic career, avoiding poverty only because his brother Eliab managed his affairs.

Leaving school at the age of 15 he went up to Caius College, Cambridge, a college which had the special privilege of each year being able to appropriate for its own use the bodies of two executed criminals. Harvey won a scholarship with an annual worth of £3 8d. It ran for six years, thus enabling him to go to Padua in Italy and study medicine. This school was then at the apex of medical teaching in Europe, and young Harvey was able to soak up the influence left by the legendary Andreas Vesalius as well as other luminaries such as Fallopio and Fabricius, who are still remembered as having lent their names to bits of the human anatomy. He graduated in 1602 and returned to England.

A swarthy and testy man, Harvey invariably wore a dagger, but despite his rough edges he gained a wide following as a physician over the next few years, and was on the staff of St Bartholomew's Hospital for 37 years. He became court physician to James I and later Charles I.

His fame was such that he was occasionally called upon to be the final arbiter in odd medical cases. For example, he was asked to examine a couple of the famous Lancashire witches for skin blemishes from which they could suckle their supposed "familiar" or supernatural spirit. He made short shift of that nonsense. He also was called upon to do a postmortem on Thomas Parr, allegedly 153 when he died and whose portrait still hangs in the Ashmolian Museum in Oxford. Not to be drawn too far into the controvery, Harvey merely observed that the organs looked "remarkably healthy".

King Charles took a great interest in Harvey's experiments and placed at his disposal the royal deer in Richmond Park to be used in his groundbreaking work. For his part, Harvey was in charge of the King's two young sons at the Battle of Edgehill in 1642. Later in the Civil War, parliamentary troops ransacked his house in Whitehall and many of his clinical and anatomical records were destroyed.

Harvey married but never had children, he and his wife doting for many years on a parrot. At least here he showed human failings, for though he always considered the bird to be male, when it died a postmortem showed an egg in the oviduct.

The appointment which brought Harvey his greatest renown came in 1616 when he was elevated to Lumleian Lecturer in Anatomy and Surgery to the College of Physicians. This legacy provided for two lectures a week throughout the year for six years, by which time it was reckoned that the subject had been covered;

I am sure it had. The regulations also stipulated that each year the lecturer was "to dissect all the body of man for five days together, before and after dinner; if the bodies last so long without annoy". His preserved notes show that it was during these lectures that he formulated the idea of circulation, though they were not published until 12 years later.

When his book did appear his private practice fell away as his patients did not favour unorthodox views. However, he advanced in the College of Physicians, though due to infirmity (he had gout) was never the president.

To demonstrate his versatility, Harvey wrote several books on aspects of medicine, including in 1651 his second greatest work, *De Generatione Animalium*, regarded as the first original book on midwifery by an English author. It was to be his last publication.

William Harvey died of a cerebral haemorrhage on 3 June 1657 at the age of 79. He was buried uncoffined but "lapped in lead" at the tiny village of Hempstead in Essex near Cambridge (famous also for being the birthplace of Dick Turpin, the renowned highwayman).

His body was exhumed in 1833, and a special Harvey Chapel was constructed in the small Hempstead church, where he was reburied in a proper coffin. Despite his pre-eminence he is still there, deep in rural England, rarely visited and lying beneath a stone effigy said to be the best extant likeness of this towering medical genius.

(JL)

Australian medical women trod a stony path

The contribution of the ... Drs [Clara and Constance] Stone to the initial group of medical women and to the health of Melbourne's poor was inestimable.

AUSTRALIAN DICTIONARY OF BIOGRAPHY

The recent film *Her Majesty, Mrs Brown* showed the softer side of Queen Victoria, the human heart underneath her stiff, public image. But hers was a conservative reign. The Queen denounced "this mad wicked folly of women's rights". The idea of women aspiring to university she found not at all amusing; as for women doctors, well really ...

During Victoria's reign, women in Australia who wanted to become doctors kept bruising themselves not on glass ceilings but on concrete barriers.

In 1865, an American woman, Winifred Ferguson, arrived as ship's doctor on a freighter and applied for registration to the Victorian Medical Board. The *Medical Journal of Australia* spoke up: "There is little fear that in any British community medical women will exist as a class. They will occasionally be imported, like other curiosities, and the public will wonder at them, just as it wonders at dancing dogs, fat boys and bearded ladies."

Exit Winifred Ferguson.

In 1880, doctors were among the crowd at Melbourne gaol to see Ned Kelly hanged. The same year, Melbourne University admitted some female students for the first time. But it refused to admit Constance Stone to study medicine, so she graduated from the Women's Medical College in Pennsylvania, USA, then gained first-class honours at the University of Trinity College, Canada. After two more years at London's New Hospital for Women, now aged 34, she returned home. In 1890, Dr Constance Stone became the first woman registered as a medico in Australia by the Medical Board of Victoria.

The next year, again in Melbourne, the first women doctors qualified from an Australian university: Clara Stone (Constance's younger sister) and Margaret Whyte (who topped the honours list). Winning admission to the medical faculty was one battle, becoming a resident doctor was another. The Melbourne Hospital rejected Margaret's application for a residency; it was their loss when she became the first resident at the Women's Hospital.

In 1896, the Melbourne Hospital was to appoint six residents, but no women. The top graduates included Dr Alfreda Gamble and Dr Janet Greig, who both fought for their appointments. Finally, and with poor grace, the authorities backed down. When Drs Gamble and Greig became the first female residents in an Australian general public hospital, a magazine celebrated:

If you've been on the ramble
And broken your leig,
'Twill be fixed by Miss Gamble
or set by Miss Greig.

They never looked back. Dr Greig became the first accredited anaesthetist at an Australian hospital and later a member of the Royal Australian College of Physicians.

At least in Melbourne, the barriers were falling. Now women could study medicine, graduate, and even become hospital residents. But for many years, only the Women's Hospital appointed women to the honorary (visiting) staff.

What drove this first generation of medical women to set up a new hospital for female doctors to care for women? The medical women could not stomach the way their male colleagues put down poor female hospital patients. In 1995, Dr Shirley Roberts wrote: "...the medical staffs of Melbourne's overcrowded public hospitals were less concerned than ever about the finer feelings of destitute patients. Women ... were treated with such disregard for their modesty that many could not bring themselves to make a second visit."

Caring for her affluent private patients in Collins Street was not enough to fulfil Dr Constance. She also spent a day each week at the free "mission dispensary" in Collingwood, to which over 16 000 patients flocked every year. Between them, the two Stone sisters saw 50–60 patients a session.

Constance married Egryn Jones, a gem of a man: feminist, medico and priest in the Welsh Church. In 1896, ten medical women founded the Victorian Medical Women's Society, with Clara as the first president. They also set up medical services for poor women and their children. The Queen Victoria Hospital started as a new service, not a new building. In October 1896, it opened a free outpatient clinic three mornings a week in a hall in Latrobe Street provided by Constance's husband.

Whole families came from as far away as Dandenong or Box Hill. There were women to whom "the idea [of checkups during pregnancy] and the possibility alike were new". Women with diabetes or tuberculosis followed one another "among every other ailment and disease in an unending series".

The doctors interviewed patients in the vestry; one doctor sat at each end of the table with a plate for contributions between them. Then the doctors washed medicine bottles and dispensed their own scripts in a cupboard one metre wide and two metres tall.

Eventually, these services grew into the Queen Victoria Memorial Hospital, where women doctors treated women patients. During the First World War, the Inspector of Charities reported: "This hospital does very good work, is well and economically managed, and is worthy of better support."

With the returning troop ships, pneumonic ("Spanish") influenza reached Australia. By early 1919, the Queen Victoria became an emergency hospital. Dr Constance Stone would have been proud if she could have seen all this. But she had died of tuberculosis in 1902 aged only 46.

(GB)

EDWARD JENNER, LADY MARY WORTLEY MONTAGU AND SMALLPOX VACCINATION

It is just over 200 years since what is regarded as one of the most famous ideas to emerge from general practice was suggested. In May 1796 an English country GP, Edward Jenner (1749–1823), following his observation that milkmaids who suffered from cowpox enjoyed immunity to smallpox, came up with the fancy that vaccination with a comparatively mild malady would prevent infection from what was one of the most deadly diseases of the era.

Jenner is held in high regard, almost reverential awe, by the medical profession for this single influential piece of deduction. But was he really the first?

In fact, the procedure has a long lineage. The ancient Chinese are said to have inoculated against smallpox by inhaling a snuff made from the dried pus of the scabs. For generations old women in Turkey had scratched smallpox pus into the arms of friends and family, alleging a minor dose now would abort a major attack later. We know a great deal about smallpox in Turkey because early in the eighteenth century, nearly 100 years before Jenner, the British ambassador's wife attended gatherings involving the dissemination of infected pus and recorded the details. She took a particular interest because at the age of 26 she herself had been disfigured by smallpox and her younger brother had died of the disease.

More than that, the lady in question, Lady Mary Wortley Montagu (1689–1762), was no ordinary wifely appendage about the embassy. She was resolute, intelligent, articulate, flamboyant, one of the first great women travellers of the Orient and altogether a force to be reckoned with. Daughter of the Earl of Kingston, Mary was born in 1689. As a teenager, rather than go through with an arranged marriage, she eloped with Edward Wortley Montagu, a member of parliament. Despite the scandal, in 1716 he was appointed ambassador to Turkey, and they went to live in Constantinople.

On 1 April 1717 Mary wrote to a friend:

I am going to tell you a thing that I imagine will make you wish you were here. The smallpox ... is here, entirely harmless by the invention of grafting. Old women ... perform the operation in September when the heat has abated. People send to know if any of their families have a mind to have smallpox. They make parties (commonly 15 or 16 together) ... the old woman comes with a nutshell full of matter ... asks what vein you are pleased to have open. She rips it open with a needle ... puts in as much venom as can lie on the end of a needle.

The children are in perfect health until the eighth day, the fever then seizes them ... in bed 2 days. About 20 pock marks ... on the face. They take the smallpox here by way of a diversion, as they take the waters in other countries.

Reading about it now the procedure seems to have been a very chancy exercise, the aim of which was to induce a mild dose of the malady so as to confer lifelong protection without pock marking. But how mild is mild?

On the diplomat's return in 1718 the couple lived in Twickenham where the marriage began to fall apart. But Lady Mary was enthusiastic in her desire to introduce the so-called "variolation" prophylactic method into England. The disease had killed one-fourteenth of the population of London during a 42-year period before 1723, and during bad epidemics up to 40% of those stricken had died. One of the snags of variolation was that, as often as not, the recipient contracted the full-blown disease, not infrequently fatally.

Several members of the British royal family tried it, as did several other Continental royal families including that of Catherine the Great in Russia. Indeed, the English surgeon who did their minor operation was awarded £10 000 and a Russian barony. Mercifully, there were no dire effects from the procedure. However, trying to enlighten her countrymen proved to be a thankless task and Mary was hooted at in the street and roundly condemned by Church and Parliament. Her revolutionary ideas about the prevention of smallpox faded, but Lady Mary continued to write with brilliance and versatility until she died in 1762.

Was she ignored and vilified because the idea of vaccination came from what was regarded as a socially different, and thus medically suspect, country? Or was it because it did not develop through the usual medical collegiate channel? Or could it be that the idea was brought into the country by a woman, and therefore regarded by the dominent male society with reservations from the start? Whatever it was, today Lady Mary Wortley Montagu is discounted as a medical innovator and is regarded merely as a colourful, eccentric, observant traveller, author and early feminist, while it is the name of Edward Jenner that is bracketed with smallpox and its prevention.

Born in 1749, Jenner was the ninth child of the vicar of Berkeley, a village in the pretty West Country county of Gloucestershire, near the Cotswolds. Orphaned at the age of five, he went to school at Cirencester where, along with being purged and bled for various childhood ills, he was inoculated with pus from a smallpox sufferer, or variolation, à la Lady Mary Wortley Montagu. Jenner got the disease but mercifully recovered.

The boy developed a keen interest in natural history while at school, a diversion which persisted throughout his life.

At the age of 14 Jenner became apprenticed to Mr Ludlow, a surgeon in the nearby town of Chipping Sodbury. One day a milkmaid presented with a rash. While pondering the diagnosis she interjected with what became the immortal words, "Well, whatever it is, it can't be smallpox because I've had cowpox, and no-one who gets that ever gets smallpox". This verity stuck in young Edward's mind.

After seven years' apprenticeship, the 21-year-old went on to St George's Hospital, London, as the junior resident of the famous anatomist and surgeon, John Hunter. While there, and knowing his interest in nature, Joseph Banks invited Jenner to go to the Antipodes as a botanist. Jenner refused, which was a pity for the medical history of Australia.

In 1772, the young graduate returned to Berkeley as the village GP. His small workload allowed him to pursue his interest in natural history, a course encouraged by Hunter who, when asked by Jenner about the lifestyle of hedgehogs, proffered his immortal advice, "Why think, why not try the experiment?"

Jenner's first publication came out in 1783. Its title, *Observations on Emetic Tarter*, gave no hint of the riches to come. The breakthrough came in 1788 in a paper read to the Royal Society. In it he described for the first time ever how the cuckoo ejects the rightful eggs from the nest. Conclusive photographs of this unique act were not forthcoming until 1929, but the observations earned Jenner a prestigious Fellowship of the Royal Society. The same year he presented a paper at the Gloucestershire Medical Society which contained the first description of mitral stenosis (a narrowing of the mitral valve in the heart) and its relationship to rheumatic heart disease. This is now well recognised.

All this time he lived and practised at Berkeley in a commodious house he called the Chantry. Mindful of the milkmaid's comments of years before, he collected cases of cowpox, some of whom he challenged with variolation. None contracted smallpox. Finally, on 14 May 1796, he saw milkmaid Sarah Nelmes who presented as a classic case of cowpox caught from her cow, Blossom. Jenner's burning-bush moment had arrived. Taking discharge from her lesions, he inoculated eight-year-old James Phipps with cowpox. Two months later he scratched smallpox pus into the boy's arm. Nothing untoward happened. Frustratingly for the experimenter, there was no more cowpox in the village for two years. When it did recur, he inoculated a child from the cow. From the resulting lesion, Jenner transferred the material to a second patient and from him to several more through three generations. None subsequently caught smallpox. As he was not inoculating the actual smallpox germ without first protecting the subject with cowpox, the full-blown cases and significant mortality associated with variolation, where smallpox pus alone is used, did not occur.

He published his observations in 1798; the book became an instant bestseller. He dedicated the second edition in 1799 to George III and presented a copy to the King in person.

Jenner had his detractors. Gillray, the famous cartoonist, portrayed vaccinated people with cow's heads sprouting from their noses and arms. A Dr Moseley from Chelsea predicted a new race of minotaurs. But the majority were believers and much of his time became taken up with answering questions from all over the world. Dried vaccine was sent abroad in a quill or on linen threads.

Parliament made him a grant of £10 000 (less £900 in tax!), and he received many international honours. "But honours won't buy mutton," as he once wrote, so in 1802, he put up his plate in London. He hated it so he returned to the country where he was to spend the rest of his life, sustained no doubt by a further grant of £20 000 from a grateful government.

On 24 January 1823 he had a stroke and died two days later. He is buried in Berkeley Church, next door to the Chantry. The sorrowing locals went somewhat overboard on the wall plaque which records "... Immortal Jenner, whose gigantic mind brought life and health to more than half mankind. Let rescued infancy ... lisp out blessings to his honoured name." But I suppose their hearts were in the right place.

The village of Berkeley, on the Bristol Channel and near Slimbridge, the late Peter Scott's famous bird sanctuary, seems to have altered little since the days of Jenner. His Georgian house, four square and sensible, is now a museum. Until 1983 the display was in the nearby house which Jenner had given to James Phipps when he became gardener to the doctor. That year the Jenner Trust bought the Chantry from the church, thanks mainly to a gift of £500 000 from a Japanese admirer of Jenner's, Mr Sasakawa.

In the museum can be seen the horns and some hair from Blossom. Her hide is located at St George's Hospital. The study has contemporary furniture and some original instruments. In the garden there still stands the bark hut in which Jenner vaccinated the local poor free of charge. He called it the Temple of Vaccinia.

The last recorded case of smallpox in the world occurred in Somalia on 26 October 1977. Exactly 200 years after Jenner's groundbreaking work, the WHO gave the deadline for the destruction of all remaining stocks of the inoculum — 30 June 1999. In the event, when the date came around it was thought prudent to retain several million doses, which are now stored in a Russian vault.

Lady Mary Wortley Montagu's "variolation" and Edward Jenner's "vaccination" were different in that one involved the use of pus from a smallpox victim, so fraught with danger, while the other did not, so was comparatively safe. Rightly, Jenner's method won the day, but Lady Mary was first on the scene and justly deserves more than a passing mention and a kindly thought.

(JL)

ASPIRIN: A HEADACHE TO EARLY CHEMISTS

How many drugs do you know that are truly out of this world? When the rocket Apollo landed on the moon in 1969, it carried aspirin. Aspirin is the most popular remedy of all time and probably the world's most widely used medicinal drug. Each year people in developed countries use about 100 tablets each. Australians take more than the British or Canadians.

Aspirin and related chemicals occur in the bark or leaves of many trees and shrubs. These include willow, myrtle, poplar, black haw, wintergreen and sweet birch. About 3500 years ago, the Ebers Papyrus advised ancient Egyptians with rheumatic pains to apply an extract of dried myrtle leaves. In ancient Greece, the father of medicine, Hippocrates, used poplar juice for eye ailments and willow bark for fever and childbirth. The Chinese, North American Indians and African Hottentots have all used aspirin-containing plants.

In England the Reverend Edward Stone noted the bitter taste of willow bark which reminded him of cinchona, the source of quinine. Because cinchona was so costly, he gathered half a kilo of willow bark, dried it for three months in an oven and then ground it into powder. In 1763, he told the Royal Society that taking this extract eased his ague (fever). Moreover, he had also successfully treated about 50 grateful patients similarly, but the Royal Society didn't want to know.

Forty years later, Napoleon's blockade of Europe stopped imports of cinchona. This made demand for aspirin-type substitutes boom. People started to look for other natural sources.

One small step for aspirin, one giant step for drugs!

In 1876, Dr T. MacLagan found that extract of willow helped patients with rheumatic fever. By now other doctors were using synthetic salicylic acid made by a German professor. Such synthetics quickly replaced the dearer natural compounds. But patients disliked the bitter taste and stomach upsets from salicylic acid; they said the cure was worse than the disease.

Felix Hoffman was a chemist working for Bayer Pharmaceuticals. His father had painful, crippling arthritis but could not tolerate salicylic acid. Felix thought a related compound (acetylsalicylic acid) might be more acceptable. After animal tests came a year of successful human trials. The Patents Office registered *aspirin* as a Bayer trademark.

Within three years, 160 scientific studies on aspirin appeared. Patients recommended it to ease pain in arthritis, rheumatic fever, period pains, gout and cancer. The leading Czech novelist Franz Kafka found aspirin could even ease his unbearable pain of being.

Originally sold as a powder, aspirin became the first major medicine in tablet form. Sales boomed. Bayer reaped large profits, though Hoffman himself apparently missed out.

Until 1914, Australians took aspirin imported from Germany. Then the British government offered £20 000 for a new way to produce aspirin. To this reward, Australian Attorney-General Billy Hughes added £5 000. Melbourne pharmacist George Nicholas, who had lost most of his eyesight in an explosion with ether, joined Henry Shmith to take up the challenge. To purify their aspirin, they used kerosene tins and Mrs Nicholas's kitchen gear, caused another explosion and nearly killed themselves.

They sold the new product as *Aspro* (Nichol*as pro*duct). It became the biggest selling aspirin outside the USA.

More recently we have found other important uses for aspirin. Apart from its effect on pain and fever, doctors use it to prevent recurrent strokes and heart attacks, and to reduce the risk of developing bowel cancer. The aspirin story rolls on and on.

(GB)

JOHN SNOW, CHOLERA, AND THE BROAD STREET PUMP

Between 1947 and 1953 I was a medical student in Liverpool, England, and in 1952 I did my student obstetrical residency at the Liverpool Maternity Hospital. Between this hospital and the Myrtle Street Children's Hospital 200 metres away there was a vacant piece of land. It was about two hectares of derelict ground, chest high in weeds and littered with rusty old bicycle or pram frames and the like, and surrounded by a fence high enough to keep out the common gaze. We were near the city centre, you understand, so it must have been very valuable.

The cafe for the lower social orders of the hospital's medical hierarchy, people such as we medical students and probationer nurses, was on the top floor of the hospital, five storeys up. Actually, we all enjoyed the arrangements and appreciated being thus thrown together. One day while gazing from one of the elevated windows, I observed that earthmoving equipment was digging up the vacant ground next door. As they went about their destructive way, to our great surprise some long rotting boxes were unearthed together with leg bones, vertebrae, skulls and other skeletal bits and pieces. As we watched they were whisked away with guilty urgency.

What in fact we were looking at was a mass grave of the great cholera epidemic of 1832. It had remained hallowed ground for 120 years but was now considered by the city fathers to be far too valuable to remain an overgrown repository for any useless detritus the locals cared to throw onto it.

All was cleared and on the spot now stands, perhaps not inappropriately, a huge waste disposal incinerator. I dare not think what happened to the bones.

Twenty-five years after this incident I went to India and visited Varanasi, or Benares as it used to be called. Opposite Clarkes Hotel where I stayed on the outskirts of the city is an Anglican church, a legacy of the days of the British Raj. I went in. In the hubbub of India it is a quiet, cool, sequestered place. But the thing which struck me was that the walls are completely lined with plaques bearing the names of soldiers who had died in that hot and dusty outpost of empire 150 or more years ago. Most were in their late teens or early twenties and the brass plates record with gut-wrenching poignancy that the vast majority had died not of gunshot wounds, as you would think, but of cholera.

So in these two diverse, but, I suppose, moving episodes, I saw with my own eyes evidence of a scourge in the last century the cause of which was then unknown but the control of which was ripe for a rigorous policy of public health.

So what happened?

Well, in England at least, came the moment, came the man. He was Dr John Snow (1813—1858) and if ever there was a man who worked for the good of the public wellbeing, John Snow was he. It is not heart transplants or other clinical heroics, but public health measures initiated by the likes of Dr Snow which have been medicine's greatest contribution to the common good. The story of how he stopped a cholera epidemic by removing the handle of the Broad Street pump in London in 1854 has now passed into both folk and scientific lore. So let's see how it all came about and then revisit the old sites of his dramatic actions and look at how they have stood up to the test of time.

Cholera is a water-borne disease caused by a comma-shaped bacillus, *Vibrio Cholera*. It is recorded from antiquity. For instance, there are Sanskrit writings from 2500 years ago of epidemic diarrhoea and vomiting with the faces of victims haggard from dehydration. There are stories of it in the Ebers Papyrus of ancient Egypt and in the writings of Hippocrates, the father of medicine, who lived on the Aegean island of Cos about 350 BC.

But it was mainly a malady of the Indian subcontinent, with flurries in Europe in medieval times. It especially flourished where pilgrims gathered together; in Varanasi, for instance. The infected devotees bathed upstream and the supplicants drank the water downstream. Its impact was not just the diarrhoea and vomiting, which in itself was rapid, frightening and dramatic, but lay in the severe dehydration, a symptom which caused victims to become almost unrecognisable and appearing more like wizened caricatures of themselves.

Although the incubation period is five days, once florid it kills within hours, the subject dying of circulatory collapse and shock to the accompaniment of abdominal pain and superficial capillary bleeding which gives a blackened or bluish appearance. It not unnaturally causes a sense of hopelessness and despair in both the observer and observed, and altogether is less than an edifying sight.

Never a tremendous problem in Europe, the condition seemed to disappear completely from that continent in about 1500. It lingered on in India, however, and there was a particularly severe outbreak in 1817 which coincided with British troops being stationed in Calcutta and along the Ganges River on which Varanasi is situated. There were many cases, and when the troops were moved first to Nepal and then Arabia the disease tagged along. When the troops reached the Middle East the disease was on virgin ground, so to speak, and went wild, especially in Mecca, again with its pilgrims.

As the natural history and cause of the malady was not then known, suitable management was not instigated and it persisted. In 1831 in Cairo it killed 13% of the population. In the same year 100 000 died in Hungary and, perhaps with some justification, the peasants turned on the doctors and nobles, blaming them for the scourge and killing them willy nilly.

Cholera seemed to leap any quarantine rules with ease and soon such measures were abandoned. Eventually, it spread across the North Sea to north-east England to make, as far as this story is concerned, its dramatic debut in Sunderland in 1831. It was then, you will be at last glad to know, that Dr John Snow made his low-key and somewhat self-effacing entry into the story.

There was an outbreak of cholera at Killingworth colliery near Sunderland where the 28-year-old Snow was a working as a young medical apprentice. He saw the deaths all around him and came up with the then heretical idea that the disease was spread via the diarrhoea it caused rather than the inhalation of "bad air" arising from the corpses. For it was this so-called "miasma" theory that was the contemporary view of the medical establishment and which, incidentally, they applied to a number of diseases.

Snow regarded the colliery as a huge privy and believed that unwashed hands, shared food and, above all, faecal contamination of the drinking water was at the bottom of it all, so to speak. As the existence of bacteria was then unknown, he had no proof; it was just a hunch. Anyway, whatever his personal opinions, he was much too far down the medical ladder to make any impact, so he kept his council. Anyway, basic hygienic methods were considered a private matter and to try to change things was seen as an assault on individual liberty.

Other centres were affected, including Liverpool, as we have seen. Characteristically, the epidemic eventually ran its course and by about 1838 things had settled down.

The following year Dr Snow moved away from the rather damp and depressing north-east of the country into upmarket Golden Square in Soho, London, where he set up as a general practitioner. He was a progressive thinker and an enterprising man who later specialised in the new-fangled science of anaesthetics. He became so adept at administering these that he gave Queen Victoria chloroform during the birth of her eighth child, Prince Leopold, in 1853. But that was a bit in the future.

In 1848–49 there was an outbreak of cholera in London which claimed 11 000 lives: no small event. Recalling his former thoughts, Snow viewed with some consternation the general lack of hygiene and squalor in which his near neighbours lived. It looked uncomfortably like the fetid coalmine of his youth. He could understand the spread of the disease within a family, but how did it get to other households, even the well-to-do? He again felt that somehow the infection was spread by faecal matter getting into the drinking water. But again, any theories he had he kept to himself.

Six years later, however, he got his chance and he held back no longer.

In 1854 there was an explosive outbreak of cholera right in the region of Golden Square where Snow lived and worked. Within 10 days there were 500 deaths in his general area. Over three days in early September there were 83 casualties, 73 of whom lived close by. Significantly, he observed that they had all used the same street pump to get their household water. Eight lived further away but came to use that pump as they considered it the sweetest water around. That meant 81 out of 83 had drawn water from the nearby, and soon to be world famous, communal Broad Street pump. He knew in his bones the cause lay in the water outlet. He put the other two deaths down to ordinary mortality.

From 1582 the London water supply had been pumped from the Thames. In 1619 water pipes were laid to street pumps and some private houses. As can be imagined, the water was heavily polluted and by the nineteenth century this was a matter of concern. By then the supply came downstream from a sewage outlet, which received effluvia from not only households but abattoirs, hospitals, tanneries and other noxious sources. There were eight separate water companies on the Thames and only one had any filtration system. The already highly polluted water then coursed from the river via cracked pipes which, in turn, enabled insult to be added to injury by running close to overflowing cesspools. Snow, however, did not know this at the time, but it came out in his subsequent investigations.

Remembering Killingworth and his feeling about contaminated water, Snow demanded a meeting regarding the water supply with the Board of Guardians of St James Parish in which Broad Street lay. His ideas were heretical, so, as you can imagine, there was acrimony, recriminations and raised voices. This was the era of putting boys up chimneys to sweep them and sending children down the mines to drag the trucks while they were bent double, so polluted water, which in its ordinary state looked fine, was not high on anyone's agenda.

Except, that is, Dr Snow's, and he was not a man to be trifled with. He stood his ground. When the Board eventually asked him to make a practical and certainly inexpensive suggestion, the blindingly simple reply has gone down into medical folklore, "Take the damned handle off the Broad Street pump."

So they did and no further cases occurred.

Having said that, Dr Snow was in fact a trifle fortunate with his inspired guess, because not until 1883, about 40 years later, was the causative organism eventually found by the German bacteriologist Robert Koch, who proved for all time that the ambient noisome smells had nothing to do with it.

Flushed with success, as it were, Snow went on to examine distant pipes and found cracks and overflowing cesspools in many areas of London. With his newfound standing, he forced other parishes to do repairs.

John Snow was a man of his times, for the nineteenth century saw the first stirrings of consciousness as regards the public good. As well as Snow, Edwin Chadwick, though an irascible, arrogant man and sensitive to criticism, made a great impact on the government of the day to improve sanitation. He has never caught the public imagination like Snow, clean water and sewage disposal never do, it is unglamerous and not spoken of in polite society, and we like our heros to be paragons of good manners and preferably good-looking and athletic; Chadwick had none of these virtues.

Other contemporary social reformers included Elizabeth Fry with her penal improvements, Lord Shaftesbury who stopped child labour and Charles Dickens, the influential novelist, who stirred people with his writings of the disadvantaged and down-at-heel.

There was no hospital specifically for children in England until 1852 when the The Hospital for Sick Children, Great Ormond Street, was opened largely with money raised by Dickens. He so stimulated major philanthropic activities that one meritorious matron, Mrs Grace Kimming, squared her troubled conscience by forming a charity which she called "The League for Brave Poor Things". Although the name was nauseatingly patronising, I suppose she meant well. It is still there in the Sussex countryside, now mercifully renamed Chailey Heritage.

All these dramas of Golden Square and the great Broad Street pump saga took place nearly 150 years ago, so what are Golden Square and its environs like now? Is John Snow still remembered? In 1996 I went to find out.

Golden Square is still there in Soho, all right, not far from Regent Street with its upmarket shops, Carnaby Street with its trendy gear and Shaftesbury Avenue with its theatres. It is a smallish square by London standards, with a fenced-off grassy centre, obviously a favourite al fresco luncheon spot for local office workers.

In the centre is a statue of King George II, rather incongruously dressed in a Roman toga. A notice at the gate indicates that only three of the original nineteenth-century houses are still standing, and that one of the characters from Dickens's *Nicholas Nickelby* had lived there. Of John Snow, there is not a word.

A couple of buildings carry the famous blue plaques of the London Historical Society. Both are of medical interest. One marks the site of the country's first Throat Hospital, founded by Morrel Mackenzie in 1865, and, tucked in a corner, the other shows the location of the house of John Hunter, the eighteenth-century surgeon and anatomist. An enquiry about Snow at a couple of shops brought blank stares and a complete ignorance of the whereabouts of Broad Street.

Disappointed, I left the square via Lexington Street. Two hundred metres or so down the road it formed a T-junction with Broadwick Street. And there, right on the corner, garlanded with hanging baskets and bustling with customers, was a small pub called (surprise, surprise) "The John Snow". After the pump fiasco, it seems Broad Street had been renamed. They knew all about Snow here, all right. In fact not only is the pub sign a painting of the great man himself, but they do a brisk trade in John Snow ephemera. Upstairs there is a collage of contemporary newspaper and magazine clippings of the famous events, plus another fine picture of our hero.

"Always getting foreigners in here," the landlord assured me. "Where did you say you were from, sir?" I told him again, trying to disguise my English accent. "Marvellous! Here, sign the book." The visitors book was full of the names of other dewy-eyed medicos on a similar pilgrimage. They came from America, Chile, France, in fact all over the world. I added my name with some pride. Even though I knew my figure is not cut out for such fol-de-rols, I was shamed into buying a John Snow T-shirt with his story emblazoned on the back.

"But what of the pump," I enquired of my host. Diagonally across the road was pointed out the replica set up by the local council within the last 20 years. It looked innocent enough, and I noted with some satisfaction that it had no handle. The inevitable plaque told the story yet again. But the landlord pulled me aside and I was aware that he had dropped his voice and was stooping to get closer to my ear. "Never mind that rubbish, this is what gentlemen like you want to see."

He took me outside where, directly under the swinging painting of Snow, he pointed to the edge of the road. What I saw was a granite kerbstone, twice the size of its neighbours and red rather than grey. Its irregularity made it look strangely out of place. The publican bent close and took my elbow, perhaps thinking I was going to swoon from the enormity of the moment. "That," he hissed conspiratorially, "that is the very site of the pump."

And so it was. The plaque on the wall confirmed the fact.

Although the stone itself was anything but inspiring, I felt oddly moved by what the site and story had come to represent to so many people, and I could not help but remember the words of the great Pericles when in 429 BC he wrote:

For the whole world is a sepulchre of famous men,
And their story is not graven on stone over their native earth,
But lives far away without visible symbol,
Woven into the stuff of other men's lives.

I went away contented.

(JL)

Chapter 2

FRAUDS

DOCTOR SUMMERLIN
AND HIS MICE

As the elevator carried him towards the director's office, it propelled him towards imminent extinction. The problem concerned the mice, which were the wrong colour.

Summerlin coloured in the patches of transplanted skin on the white mice. In an instant he had succeeded in producing the vital skin graft that years of research had not yielded.

YOUNGSON & SCHOTT, *MEDICAL BLUNDERS*

In 1972, American President Richard Nixon declared a $1.7 billion research war on cancer.

Immunologist Dr Robert Good was Director of New York's elite Sloan-Kettering Institute for Cancer Research. Still only in his fifties, his photo appeared on the cover of *Time* magazine. Then 35-year-old Dr William Summerlin, tall, charming and persuasive, came to work under Dr Good. These two looked like setting the fiercely competitive world of cancer research on fire.

The *Time* article on Good also promoted Summerlin as having made:

a discovery that may well make tissue typing unnecessary. [He has] found that when skin is kept in tissue culture for several weeks, its antigens (chemicals that stimulate an immune response and so cause rejection after organ transplantation) are somehow lost. The skin can then be grafted on to any patient without being rejected. Summerlin's work . . . could eventually eliminate both the rejection problem and the need to match donor and recipient, enabling transplant surgeons to make wider use of organs taken from cadavers (corpses).

Summerlin certainly looked like a young man with a great future. But only a year later, on 26 March 1974, he not only destroyed his own career but also made waves that affected the credibility of medical research generally. He did this by drawing black patches onto two of the mice in the cage he was about to show Dr Good. Summerlin had been culturing and then trying to graft black skin onto white mice. By darkening the colour of the donor areas, he made it appear that his grafts had taken.

How did such a promising rooster turn so quickly into a featherduster?

During the peak of the Vietnam War (from 1965) Summerlin worked at a hospital burns centre in Texas. The problem of covering large burns stimulated him to turn from surgery to research.

A black man had a third-degree burn of about 10 cm on his flank. Summerlin set up cultures of pieces of skin from a live white donor. After a week of culture, he tried to graft the white skin, but the patient rejected it. After two weeks, the same thing happened. After three weeks, the white skin lasted longer. After about four weeks of culture, the black man was no longer rejecting the white skin.

**Dr Summerlin's story regarding his
mouse skin-grafting research sounded
very silly to anyone who cared to listen.**

Turning to mice, Summerlin reportedly found that culturing skin or other organs for some weeks before transplantation made them immunologically neutral (so they no longer caused rejection). He had similar results with corneas from eyes. His technique, he claimed, even allowed him to transplant from one species to another! That was unheard of.

But other workers could not confirm his results. Despite his high media profile, Summerlin faced growing scepticism among his peers. In fact, a junior member of Summerlin's own team was pressing to publish a paper announcing his own failure to reproduce Summerlin's transplant successes.

At 4 a.m. on 26 March 1974, Summerlin rose from the cot on which he often slept in his office. It was on the way to his 7 a.m. meeting with Dr Good that he inked the black patches onto the scars on the white mice that he had grafted with skin from black mice. If Summerlin had only known how futile his clumsy fraud was. Dr Good hardly looked at the mice. It was the lab assistant to whom Summerlin returned the mice who found that alcohol washed away the "skin grafts". Once he reported it, Summerlin's superiors suspended him.

Dr Good reproached his protege: "We could have gone together to Stockholm [to receive the Nobel Prize]. But now you are dead, dead, dead."

They allowed him up to one year's leave on full pay "to enable him to obtain the rest and professional [read psychiatric] care which his condition may require." It suited his superiors to imply that Summerlin had gone out of his mind. But a committee also blamed Dr Good himself for allowing Summerlin to publicise his findings so widely before they had been confirmed. "Good was slow to respond to a suggestion of dishonesty against Dr Summerlin ... when several investigators were experiencing great difficulty in repeating [his] experiments."

How could Summerlin expect such a gross scam to succeed? A colleague said: "The man had everything going for him. Then he threw it away; it's almost like Rembrandt putting his foot through one of his paintings." In his own defence, Summerlin spoke of the extreme pressure to produce results and so support applications for research grants. Robert Youngson and Ian Schott sum up: "While Summerlin was principally to blame, he was very much the product of a research environment that required, for prestige and funding from the drug companies, continuous new discoveries."

Before this scandal broke, idealists liked to believe that scientific fraud was rare. How many cases remain undetected?

(GB)

Chapter 3

Versatile doctors

Spies, lies and secrets

In 1954, Vladimir Petrov, the senior Soviet intelligence officer in Australia, defected to ASIO (Australian Security Intelligence Organisation), taking with him a bag stuffed with documents. His defection led to intense political controversy, a Royal Commission, the re-election of the Menzies Liberal Government, a split in the Australian Labor Party (ALP) and the political death of Dr H. V. Evatt, leader of the ALP.

Dr Michael Bialoguski, the man behind the Petrov defection, was born in 1917 to Polish parents in the Ukranian capital, Kiev. How did this man have such an effect on a country so far away?

By 1920, Kiev had been a First World War and Russian Civil War battleground occupied by the Bolsheviks, Germans, White Russians and Polish–Ukranian forces. Bolsheviks were about to shoot three-year-old Michael and his family, when Michael's father managed to bribe them with his gold watch. The family had to flee on the spot. They wound up in the Lithuanian capital Vilnius, then occupied by Poland. As he grew up there, Michael learned to play the violin and in 1935, he started medical school.

When the Germans and Russians invaded Poland in 1939, police arrested Michael for possessing firearms. Released after three months in gaol, he led the orchestra of a musical comedy troupe, laundered money, worked the blackmarket and endured another spell in gaol. His first marriage failed.

In early 1941, carrying a one-carat diamond hidden in a toothpaste tube, Michael set off on the Trans-Siberian Railway and crossed the whole Soviet Union (about 10 000 kilometres) to Vladivostok on the Pacific Coast. Here he used his fluent Russian to talk his way onto a ship bound for Japan. In Tokyo, he sold his diamond and spent some weeks wangling the documents to migrate to Australia.

When Michael walked alone down the gangplank in Sydney in June 1941, he spoke no English, knew no-one and owned only a violin, a few shirts and £13. At first he supported himself playing the violin at concerts and on the radio. Within a year, he had served briefly in the Auxiliary Medical Corps and resumed his medical studies (paid for by the Repatriation Department). His good looks, smooth Continental manner and strong personality won him many hearts, including that of Patricia Humphrey, who divorced her husband to marry him.

In 1945, with the Red Army spreading across Eastern Europe, the Commonwealth Investigation Service (CIS) engaged Michael to "unmask foreign agents". To find these subversives, he joined the NSW Peace Council and the pro-Bolshevik Russian Social Club. Later he moved to ASIO, where he earned £4 a week. At his general practice in Macquarie Street, Sydney, he treated many refugees from the Soviet Union and also did illegal abortions. Sir Eugene Goossens persuaded him to play first violin with the Sydney Symphony Orchestra.

In 1951 at the Russian Social Club, Michael met Vladimir Petrov, who had joined the Soviet Embassy in Canberra. Though he was nominally Third Secretary, ASIO suspected Petrov of being a top intelligence officer.

So started a bizarre courtship. For almost three years, the Polish spy–doctor and the Russian spy–diplomat circled, manipulated and exploited each other. Each spent more energy on this relationship than on his own marriage. Whenever Petrov visited Sydney, they would cruise the town in search of prostitutes and drink. As Petrov slept off the drink, Michael emptied his pockets and copied the contents. Not suspecting that Michael worked for ASIO, Petrov gave him some minor espionage tasks.

Michael insisted that Petrov was a potential defector and resented ASIO's refusal to believe him. His superiors in turn called Michael presumptuous, avaricious and unreliable; they even hired an agent to watch him!

Several times Moscow reprimanded Petrov and his wife Evdokia (also an undercover spy) on their work and on his drinking. When Moscow summoned Petrov home, he feared the worst. He complained of an eye problem, for which Michael took him to specialist Dr H. C. Beckett. By now ASIO hoped that Petrov would indeed defect but wanted to bypass Michael. So they approached Petrov through Dr Beckett, who had no intelligence background. Soon ASIO head Colonel Charles Spry dismissed Michael altogether.

Finally, after an infuriating three-year saga of coaxing, hesitations and delays, Petrov did defect. On 3 April 1954, at a safe house on Sydney's North Shore, an ASIO agent handed him £5000 in exchange for intelligence documents. He was the most senior Soviet spy to defect to the West since the 1930s. "Without this extraordinary character [Michael] there would have been no Petrov defection," wrote Robert Manne, author of *The Petrov Affair.*

On announcing Petrov's defection, Prime Minister Robert Menzies set up a Royal Commission to investigate Soviet espionage in Australia. Federal Labor Leader Dr H. V. Evatt convinced himself that the whole defection was a sham, or else timed by Menzies and ASIO to coincide with the 1954 Federal election. Evatt's outbursts against ASIO became so outlandish and grave that the Royal Commission felt obliged to let ASIO defend its honour.

Michael Bialoguski's lengthy cross-examination enabled him to speak in public about his key role in the defection.

But what drove Michael? His second wife, Patricia, said that he was fascinated by espionage: "[he was] supremely ambitious, both for wealth and fame ... self-pitying, resentful and vengeful ... prone to fits of irrational jealousy". She quoted Michael: "No law can hold me. I am my own law."

In 1957, after writing his own book on the Petrov affair in 1955, Michael married for the third time; this marriage produced three children. I heard the story in 1998 from his last wife Nonnie and their son Stefan. She described their time in Australia after the Royal Commission as hell, with both sides of politics gunning for Michael. The left reviled him as part of the Petrov conspiracy and a lackey of the conservative government. On the right, ASIO did not want to share the credit for Petrov's defection.

The family moved to England in 1964, where Michael settled into general practice but made a musical stir as well. Once he hired the Albert Hall and a symphony orchestra that he conducted before a large audience.

Dr Michael Bialoguski died in 1984. The last word about him rests with Nonnie. She admits he was arrogant, difficult and paranoid. "He was always the alien — the outsider. Perhaps he over-compensated — fighting the world. Alone!"

(GB)

DOCTORS AND BUCCANEERS

Buccaneers and doctors always enjoyed a close relationship. By the nature of their trade, pirates often had to call on medical services, however crude, and doctors have never been averse to investing their energies into business ventures, however risky.

Indeed, the word "buccaneer" was coined by a doctor. During their heyday in the second half of the seventeenth century they were usually called "privateers". But in 1684 there appeared a book called *Buccaneers of America*, written by Dutch surgeon/pirate Alexander Esquemeling. Unlike many of his fellow so-called "Brethren of the Coast", Esquemeling was articulate and well read; his book became a classic and the word itself, derived from the French *boucan*, a grill for smoking dried meat on ships at sea, passed into the language. Presumably, this form of cooking was standard among the sailors of that kind.

Alexander Esquemeling had arrived on Tortuga Island near Haiti in 1666, footloose, untrained for any gainful employment and bent on adventure: an ideal candidate for some unlawful pursuit. Lax French government officials allowed the island to be a base for pirates and he soon joined up. After three years of harsh, but doubtless lucrative seafaring, he was sold to a surgeon who taught him the rudiments of the craft.

An adept student, Esquemeling subsequently became an authority on the therapeutic worth of local plants while serving as a surgeon during many of the exploits of the great pirate, Henry Morgan, as well as other villains. From the rough and ready experiences gleaned, he wrote his book which contained a first-hand account of the piratical excesses of rape and murder. Rather incredibly, Morgan himself was later to succeed in sueing the English publishers of the book for "many false, scandalous and malicious reflections" on his life.

An interesting medical twist in the book concerns how the booty was divided after Morgan had sacked Panama City in 1671. The captain himself was allotted 100 parts of the plunder. The surgeon merely got 200 pieces of silver for the use of his medicine chest. That may not sound too bad, but when you consider that at the drum head division of loot, compensation for injuries sustained in the course of their nefarious activities included 1500 pieces of eight or 15 slaves for the loss of both legs, or 600 pieces of eight or six slaves for the loss of one hand, the surgeon came pretty cheap.

Henry Morgan himself was an outstanding leader, the pinnacle of whose career was famously leading 36 ships in the sacking of Panama. It was his swan song, for shortly afterwards he "swallowed the anchor" and became a pillar of society. He retired to Jamaica, becoming Lieutenant Governor then a judge, and was eventually knighted by Charles II!

Morgan led a strangely abstemious life for the rumbustious times in which he lived and fought and he enjoyed good health as a pirate. However, in 1688 he fell ill and was treated with dubious success by a young doctor who was to attain an even greater fame in the medical profession than had Morgan privateering on the Spanish Main. He was Hans Sloane, later Sir Hans Sloane, founder of the British Museum, the Chelsea Physic Garden, and President of both the Royal Society and the Royal College of Physicians of London. Sloane Square in fashionable London, and in recent years epicentre of the "Sloane Rangers", is named after him.

This medical man has left us an account of Morgan's last illness when the retired villain was 53. He suffered from "dropsy" due, it seems, to kidney disease and for which Sloane gave a host of poultices, clysters (enemas) and emetics. They had no effect and a wrung-out Morgan eventually sent for another doctor. But he died just the same.

Perhaps the best-known name among the ranks of buccaneer/doctors was that of the Cambridge graduate Thomas Dover, of Dover's Powder fame, a concoction which can still be found in the backroom of some old pharmacies and which was in occasional use as recently as 30 years ago. It was a mixture of opium and ipecacuanha to be used as a sedative-cum-analgesic, and especially useful if diarrhoea was present. It last appeared in the 29th edition of *The Complete Drug Reference*, the pharmacists' bible. I remember the powder well, mainly for the sweating it used to cause.

Dover (1660–1742) was a successful Bristol doctor who became a medical officer on a privateering voyage which circumnavigated the globe. Besides his powder, his main claim to fame was his part in the rescue of Alexander Selkirk from Juan Fernandez Island off the coast of Chile on 2 February 1709. This castaway had survived alone there for four and a half years, and became the role model for Daniel Defoe's book *Robinson Crusoe*, written ten years later in 1719. Dover also popularised the use of quicksilver (mercury) for syphilis, infertility and indigestion — an odd case mix and useless in them all.

Lionel Wafer was a surgeon who joined the Caribbean adventurers. He wrote of his life of looting combined with surgery but is best remembered for his account of local Indians and wildlife.

Of interest in Australia is the famous pirate William Dampier (1652–1715). Between 1678 and 1691 he was engaged in piracy off South America, during which time he reached Australia but found nothing to plunder. He reformed, and in 1699 was sent by the Admiralty to explore the Antipodes in the *Roebuck*. He reached Shark Bay off the Western Australia coast and journeyed north to name the Dampier Archipelago and Roebuck Bay. Later he briefly returned to piracy, but by then the buccaneer's halcyon days of looting, pillage and rape were just about over.

Dampier was concerned about the health of his crew. He recorded that he found it very beneficial to wash morning and night, especially if suffering from the flux (diarrhoea); that old betel nuts caused giddiness but were excellent for sore gums; that too much "penguin fruit" produced "heat or tickling in the fundament" (I'm not sure what penguin fruit is or what its effect means, but the mind boggles). He became expert in the therapeutic use of many native plants.

By about 1690 various international conflicts allowed these freebooters to become legitimate privateers in the service of their respective countries. With respectability, the romantic, though bloodthirsty age of buccaneers and their surgical hangers-on came to an end.

(JL)

Sun Yat-sen: Doctor to all his countrymen

The top physician cures the nation first, and then the people.
GUO YU, ANCIENT CHINESE BOOK OF HISTORY

Dr Sun Yat-sen made his name by freeing China from the Manchu dynasty's despotic rule. His people honoured him as "Father of the Revolution"; many revere him still as the father of modern China.

Sun Yat-sen was born in 1866 to a Christian peasant family in a village in Kwangtung (Canton) province. With sweet potato as their staple food, the family survived on their one-acre tenanted field. At the age of 14, Sun stowed away to join his older brother in Hawaii. There at an Anglican mission school, he became so fond of English ways that his brother shipped him back home. At the age of 18, as was the custom, he entered an arranged marriage. Next he moved to Hong Kong, where he very nearly became a Christian missionary, but instead graduated as a doctor. What, then, made him take up the life of a reformer and revolutionary instead? Perhaps it was outrage at the slave system in China, under which families sold their daughters for prostitution and their sons into bondage.

Somehow Sun juggled medicine with top cricket and undercover activity. Since the Hong Kong General Medical Council did not recognise Sun's qualifications, he set up instead as a herbalist and offered both traditional Chinese and Western medicines.

From 1892, he practised in Macao (Macau), becoming the first doctor trained in Western medicine to work in a traditional Chinese hospital. There he persuaded the governors to devote one wing of the hospital to European methods, the other to Chinese practice, and then to compare results. When Sun was operating, the governors would sit near the operating table, with the patient's relatives and friends standing nearby. Sun's skill impressed them all.

He also borrowed money to set up a Chinese and Western pharmacy, where he saw patients and dispensed medicines without charge. But Portuguese physicians, resenting his success, forced Sun out of Macao.

Gradually, the reformer displaced the doctor. He pressed for political reform, universal education, improved agriculture and the prohibition of opium. But while he worked inside the existing political system, all Sun's efforts at reform remained futile. In 1894–95, defeat in the war with Japan forced China to give up Taiwan; then Britain, France, Germany and Russia grabbed more Chinese territory. Clearly, the weak, corrupt Manchu government could not protect his country. Sun set up his own activist group,

"Dare to Die". When the authorities executed his comrades, he himself was lucky to escape.

So began his lifelong succession of crises, unsuccessful plots and coups, flight from murdering Manchu agents, exile and wandering. For years on end, Sun shuttled between Hawaii, Japan, Europe and the USA, to raise money to overthrow the despotic government of China. In all, he was behind ten uprisings.

By 1896, he had a price of £100 000 on his head. While walking near Portland Place, in London, Sun was kidnapped and held for 12 days inside the Chinese legation. His captors were about to ship him back to a certain death in Beijing when a maid from the legation secretly notified Sun's old teacher Dr James Cantlie, now living in London. All that night Cantlie and Dr Patrick Manson (of malaria fame) kept vigil outside the legation until the morning when the Foreign Office, Scotland Yard and *The Times* all demanded Sun's release. Sun's international reputation soared when he published his account *Kidnapped in London*. Like Karl Marx and Vladimir Lenin before him, Sun spent months of study in the British Museum library.

A bomb explosion in Hankow in late 1911 marked the beginning of yet another revolution, but this one succeeded and the boy emperor abdicated. At a convention in Nanking, Sun was sworn in as president. But the republic had little support and an inevitable power struggle followed. Realising that he could not keep the country united under his own rule, he resigned in favour of Yuan Shih-k'ai who at first seemed to be another reformer. But Yuan soon declared himself emperor. Within a year, Sun and Yuan were leading armies against each other. The country was plunged into anarchy and terror; Sun had to flee once more and could not return until Yuan died.

Even then, North and South China were still divided; Sun worked towards an independent republic of South China, of which he became president. Unable to work with the army leaders in Canton, he resigned once more. There followed a military dictatorship. In 1921, this fell and Sun became president yet again. New alliances formed and dissolved with bewildering speed.

Since the Western democracies would not support the Chinese revolutionaries and were profiting from China's weakness, Sun felt he had to sign a pact with the Soviet Union. In 1923, Communist Russia sent money, arms and advisers to Sun's government.

While in Peking for a People's Conference, Sun came down with a raging fever. An operation showed advanced cancer; now neither Western nor Chinese medicine could help him. On 12 March 1925, at the age of 59, Sun Yat-sen died. Four years later, his body was interred in his own mausoleum in Nanking.

Sun's critics saw him as too idealistic. He sometimes planned badly, failed to act decisively and trusted generals who betrayed him. So what was his appeal? Sun had a magnetic personality, ambition for power and outstanding knowledge of the West. Politically, Sun dead was even more effective than Sun alive.

His *Three People's Principles* — socialism, democracy and nationalism (unification of the many peoples of China and an end to their exploitation by Europeans) — inspired his successors. Both nationalists and communists revered Sun as the founder of the Republic of China.

In 1946, the nationalist Chiang Kai-shek (who had been a protégé of Sun) formed a government and fleetingly achieved Sun's dream of unifying China. But the communists overthrew him. In 1949, Mao Zedong headed the People's Republic of China, forcing Chiang and his nationalists to withdraw to Taiwan. Sun's young widow, Soong Ch'ing-Ling, became a prominent figure in the Communist government.

Even now, over 70 years after his death, Sun's dream of a unified China remains just that: a dream. Even now, we cannot foretell the end of the story of modern China. What we do know is that without Dr Sun Yat-sen there would not even have been a beginning.

(GB)

DR LIVINGSTONE, I PRESUME?

One of the great, understated epigrams typical of the Victorian era was "Dr Livingstone, I presume?". It was reputed to have been said by H. M. Stanley, a journalist from the *New York Herald*, when, after 10 months trekking from the coast of East Africa to find the supposedly lost David Livingstone, he eventually confronted the doctor/missionary on 28 October 1871 at Ujiji, the Arab slave-trading centre on Lake Tanganyika.

So who was this great medical missionary-cum-explorer who worked in Africa for over 30 years, and why was he lost (if indeed he *was* lost)?

Although Livingstone's most well-known exploration was when he went down the Zambesi River to find the falls of Mosi-oa-tunya ("the smoke that thunders") which, with a jingoism typical of his times, he immediately renamed the Victoria Falls, being an explorer was definitely his third priority in life, and only seriously undertaken after his first two loves had been satisfied.

The intrepid H. M. Stanley, N.Y. journalist.　**The obviously named Dr Livingstone.**

First and foremost Livingstone (1813–1873) was a committed Christian who, again typical of the era, believed that the British were the most suitable race to convert the African "heathen", and he himself more suited than most. He saw spreading The Word as his mission, and believed that divine guidance would eventually lead him to his ultimate goal — the spot where Moses once bathed in the Nile. Such was his devotion to this primary duty that in the hope of preaching more clearly, Livingstone had his uvula removed in Cape Town in 1852. The uvula is that small piece of tissue that hangs down at the back of the throat. I hope it worked, though the effect of surgery on a Scottish brogue must have be equivocal at best.

As curing the indigent sick was seen to be an essential part of this ministry work, so it was that medicine became his second love. In the context of medical history, it is this aspect of his life that is of more interest to us.

David Livingstone was born in Blantyre, Scotland, in 1813 into a deeply religious family. To augment the family income, at the age of ten he left school to work in a cotton mill. By the age of 13 he was working 14 hours a day, but such was his desire to be educated that he attended classes in the evening for another three hours. During the day he propped up his textbooks on the spinning jenny.

At 20 he joined the London Missionary Society with a view to foreign pastoral work and was told that medical expertise was a highly desirable sideline. So with money from his brother, Livingstone walked from Blantyre to Glasgow to enter the Andersonian Medical College. Fees were £12 a year and his lodgings cost 2s 6d a week. During the holidays he went back to work in the mill to help pay the fees.

After two years he travelled to London to study theology, coupling it with clinical studies carried out at the Charing Cross Hospital. As he could not afford to pay the fee for the London examination, Livingstone returned to Glasgow to sit the papers and qualified Licentiate of the Faculty of Physicians and Surgeons of Glasgow in 1840. Within a week he was ordained a missionary and a fortnight later sailed for Africa and into history.

At the time the diseases of that continent were a complete mystery. Over the years Livingstone was to make many clinical observations, though he never got them together to formulate any great medical discoveries, and in the light of his nineteenth-century medical knowledge he often drew erroneous conclusions. For instance, he knew malaria was the main scourge of the continent, though its relationship with the mosquito was quite unknown. He felt that victims who kept active suffered least and advised accordingly. He also devised the "Livingstone Pill" for the condition; it contained resin, jalap, calomel, quinine and rhubarb. Doubtless the quinine, empirically given as it was, may have had some therapeutic effect; the purgative nature of the rest would at least take your mind off things.

He gleaned other odd observations from native culture, including the putting to death of a child who cut the upper before the lower incisor teeth. Another was the eating of earth, mainly among slaves and pregnant women. He could not explain this, but nowadays we would put it down to an iron deficiency within the subject.

Livingstone was among the first to note the effect of the blood-sucking tsetse fly on cattle, yet wrongly thought humans immune from any diseases these insects may carry. In fact, they are a vector in the life cycle of various maladies, especially sleeping sickness. He recorded that goitre, endemic in the highlands, disappeared within a few days of drinking water from Lake Tanganyika, which was a perceptive observation, but as the necessity of iodine water to prevent this condition was quite unknown to him, it meant nothing. He was a man before his time.

In the countryside Livingstone soon became more popular as a doctor than a man of a strange God. However small his therapeutic armamentarium, the locals thought that, like his firearms, his medicine was magic. Indeed, they were always pressing him to show them the secret of his "gun medicine", as they called it.

Over the years he himself suffered over 30 attacks of malaria, as well as many other extremely debilitating exotic tropical diseases.

He saw them as being divinely inspired tests of his devotion; he never wavered from his chosen path. But after 33 years it was probably none of these exotica that finally killed him.

Livingstone had always been troubled by haemorrhoids but refused surgical treatment when on leave, believing that the blood loss relieved his headaches (another false conclusion). However, by 1871 the loss was almost continuous. Becoming debilitated from this bleeding, he decided to make for the coast and seek help. As he made his slow and tortured way, the missionary became weaker and weaker. Attacked by driver ants, squelching through marshes, feverish, exsanguinated and emaciated, in the end his men had to carry him by litter. It was all too much for even such a hardy battler and on 1 May 1873 he died in Chitambo's village on Lake Bangweulu (Zambia).

There then occurred a most remarkable sequence of events.

Livingstone's bearers opened up the body, removed the internal organs and heart and buried them. They were replaced by salt. The corpse was then exposed to the sun for 14 days to dry, after which it was tied in calico and strips from the myonga tree. The whole body was coated with tar and strung to a pole to be carried between two men, Susi and Chuma. It took until February 1874 to reach the coast, and until April to sail to England.

With contraction and flexing of the legs, the package was only 1.2 metres in length when it arrived. A postmortem was undertaken by Sir William Fergusson, President of the Royal College of Surgeons and surgeon to Queen Victoria. He recorded that the face was unrecognisable, but that one feature clinched the identity; at some time in the past the left humerus bone in the upper arm had been fractured and now displayed an oblique disunited fracture with a false joint. The surgeon recognised it, for thirty years before Livingstone had been mauled by a lion and sustained a compound fracture of the left upper arm. It became infected and never properly healed, and, incredibly, while on leave the missionary had consulted this self-same William Fergusson about the unsatisfactory union of the bone.

The story of David Livingstone's selfless life and tragic death touched all strata of society and earned him a burial in Westminster Abbey. In part his epitaph reads:

> *Brought by faithful hands*
> *Over sea and land*
> *Here rests*
> *David Livingstone,*
> *Missionary, Traveller and Philanthropist.*

His vast impact on medicine does not get a mention. Nonetheless, to help enact the final rites there were present in the Abbey not only the redoubtable H. M. Stanley, but also Susi and Chuma, who revered his spiritual and clinical dedication enough to have carried the body over 1000 kilometres, keeping the faith for over eight months.

(JL)

Dr Livingstone, I presume?
I'd recognise that fracture
in the left upper humerous bone anywhere!

SERVETUS SHOULD HAVE STUCK TO DOCTORING

In 1628, William Harvey described the pumping action of the heart and the circulation of the blood. But Harvey, like most pioneers, was building on the work of his predecessors. As early as the second century, the Greek physician Galen saw one side of the heart contracting before the other. From this, he theorised that blood passed through tiny holes in the muscle separating the two sides of the heart. Though no-one could ever see these pores, Galen's godlike authority led physicians to accept their existence for 1400 years.

A thirteenth-century Arab, Ibn An-Nafis, described the circulation of blood from the heart through the lungs and back again, but his findings made little impact. The first European to confirm them was the Spaniard Michael Servetus (Miguel Serveto, Servede — c. 1511–1553).

Physician, physiologist, astrologer, geographer, mathematician and fearless theologian, Servetus had the breadth of a true Renaissance figure. But his contemporaries never acclaimed his genius; they shunned him as a threat to religious orthodoxy and hence to political stability.

Servetus was a precocious youth. After convent school, he probably attended the University of Saragossa, where he excelled in Latin, Greek and Hebrew. Then he went to France, to study law at Toulouse. Like other religious reformers, Servetus could not stomach the worldliness and corruption of the papacy. Nor could he accept the Church's absolute authority: "It would be easy enough, indeed, to judge dispassionately of everything, were we but suffered without molestation by the churches freely to speak our minds."

In 1531, Servetus published a text against the doctrine of the Holy Trinity (*De Trinitatis Erroribus*). He expected support from the more liberal reformers, but both Protestants and Catholics were outraged. Bucer, reputedly a kind man, said the author deserved to be disembowelled and torn in pieces!

In Paris, Servetus studied medicine. He came under the influence of "the anatomical arch-heretic" Andreas Vesalius. Unlike the followers of Galen, Vesalius learnt his anatomy not from old books but in the dissecting room, and was sure that Galen was fallible. Their spirit of inquiry, painstaking dissection and observation bore fruit in Servetus's discovery of the pulmonary circulation. Servetus wrote a medical book that contradicted Galen. He also outraged his faculty by lecturing on the forbidden topic of astrology (fortune-telling: the belief that the stars influenced health).

Despite all this, Servetus graduated as a physician in 1538. He went to practise medicine and study theology at Louvain in France. He was a good, kind doctor. Had he stayed quiet and stuck to medicine, all might have been well. Instead, his questioning and wandering brought him into perpetual conflict and finally precipitated his death. Servetus's life had the inevitability of a Greek tragedy. Few heretics managed to get so seriously offside with both camps. After the Catholics burned Servetus in effigy, he provoked the Protestants to do so in the flesh. Was he just naive? Or did he have a death wish?

Medicine was his livelihood, but theology was always his passion. By 1546, Servetus had drafted his major work *The Restitution of Christianity* (*Christianismi restitutio*). He hoped this work would purify the Church from its errors of doctrine, and so win the world to Christ.

He sent a copy to the Protestant reformer John Calvin in Geneva, asking Calvin to meet him to talk. Had he forgotten his criticisms of Calvin's major work, *The Institutes of the Christian Religion*? Calvin declared: "This limb of Satan ... if this creature ever dares to visit Geneva, he shall not ... leave alive."

In 1553, Servetus published 1000 anonymous copies of his revised *Restitution*. It was 700 pages long. Tucked away among the theology was his rediscovery of the pulmonary circulation. But it was the theological content of *Restitution* that provoked his arrest. Was it John Calvin who denounced him to the Catholic Inquisition in France?

Though tried and convicted of heresy in France, Servetus escaped from prison and planned to flee to Italy. Whatever drew him instead to Geneva? Perhaps he expected the more liberal Protestants, who opposed Calvin, to prevail there. A forlorn hope. This time it was Calvin's supporters who arrested him. Servetus's condemnation for heresy had ten headings; the two most important dealt with the Trinity and Infant Baptism.

Calvin pressed for execution but, to his credit, endorsed Servetus's request for a merciful mode of death. In vain. Within sight of beautiful Lake Geneva, they led him to the stake. Even then, he might have saved himself by recanting. To prolong his agony, they used green, moist wood. Around his waist they tied his manuscripts and his *Restitution*, of which only three copies have survived.

Sir William Osler's view: "Judged by his age, Servetus was a rank heretic, and as deserving of death as any ever tied to a stake. We can scarce call him a martyr of the Church. What Church would own him? All the same, we honour his memory as a martyr to the truth as he saw it."

(GB)

DOCTOR WAKLEY JUGGLED THREE CAREERS

Wakley pursued careers in ... three areas ... any of these would have been a full-time assignment for the average man. He was founder, owner, and editor of what has been perhaps the most influential medical journal in the world. He was a Radical politician, a member of the British parliament ... he was a combative coroner and leader of a movement to have physicians be coroners in a day when that was unheard of.

CHARLES G. ROLAND, INTRODUCTION TO S. SQUIRE SPRIGGE,
THE LIFE AND TIMES OF THOMAS WAKLEY

As a youth, Thomas Wakley (1795–1862) excelled at billiards, boxing and cricket. As an adult, he followed a more serious trifecta, becoming founding editor of the *Lancet*, a coroner and a member of parliament.

His father was a farmer in Devon. At the age of 15, Thomas became an apprentice to an apothecary and then to a surgeon. In 1815 (aged 20), he went to London. At the Royal College of Surgeons of England (RCSE), he competed with his fellow students for bodies "resurrected" from the gallows. Body snatching was still rampant. After qualifying, he used £400 from his future father-in-law to buy a West End practice, but soon took on the first of his three new careers.

Wakley the editor

In 1823, aged only 28, he launched a new medical journal, the *Lancet*. His goals were to inform and to reform. Without seeking permission, he printed verbatim reports of lectures by medical teachers. Leading surgeon Sir Astley Cooper took the public airing well. Not so his colleagues. A few London practitioners were earning large fees from the compulsory lectures they gave to students. They did not fancy seeing their teaching on sale for sixpence a week; they also resented seeing their inadequacies exposed in print.

Further, these same powerful men channelled the choice teaching posts to relatives and friends. Even Sir Astley Cooper made no bones about it when he listed the surgeons of a London hospital: "Mr Travers was my apprentice, Mr Green is my godson, Mr Tyrrell is my nephew, Mr Key is my nephew, Mr Morgan was my apprentice." Wakley wrote in the *Lancet*: "The manner in which the appointments are managed is the most nefarious ... ignorant pretenders exclude men of sound talent."

Again without permission, he printed many pointed case reports from London hospitals.

In 1825, a man was admitted to St George's Hospital with pneumonia. While bleeding him, the dresser wounded the large artery in his arm. The man was placed under one of the surgeons, who consulted a physician. They agreed that the pneumonia was too severe to operate on the artery. Instead, they applied a tight bandage to the arm above the wound. The bandage remained untouched for three days, gangrene set in and the patient died. The verdict of the coroner's jury: the man died from the "accidental opening of an artery and from the want of proper attention . . . ".

In reporting such cases, the *Lancet* fearlessly named all concerned. Over 10 years, Wakley fought ten legal actions. Except for the leniency of the libel laws, he might have spent a long time in jail or in bankruptcy. But he took in his stride the few verdicts that went against him.

To further his medical and social reforms, Wakley took on two other roles: coroner for West Middlesex and MP for Finsbury — no, not instead of editing the *Lancet*, but as well as!

Wakley the coroner

Wakley produced evidence of incompetence by coroners, which he blamed on the fact that they were not medically qualified. After being easily elected as coroner in 1839, Wakley insisted on holding inquests whenever the cause of death was in doubt.

Frederick White, a private in the Seventh Hussars, assaulted his sergeant. On 15 June 1846, he received 150 lashes; on 11 July he died. After an autopsy, three army surgeons said, "The cause of death was in nowise connected with corporal punishment." But a vicar reported the case to Wakley. After further examination of the disinterred body, the coroner's jury overturned the army verdict and ruled that the death was indeed related to the flogging.

Because of this verdict, flogging became rare, though it was not outlawed until 1881.

Wakley the MP

Soon after his election in 1835, Wakley made his mark in Parliament. He challenged the harsh penalties dealt out to six agricultural labourers in the Dorset village of Tolpuddle (the Tolpuddle Martyrs). Their punishment was transportation for seven years to New South Wales for "unlawfully administering a secret oath". Their crime was forming a trade union and protesting at the reduction in their weekly wages from seven to six shillings! By presenting 16 petitions with over 13000 signatures, Wakley won the release of all six men.

He argued that legislators must correct the causes underlying social problems. Hence Wakley deplored any move to increase penalties for infanticide. His reason? The cause of infanticide was increasing illegitimacy, which was caused by the inability of the poor to marry, which was caused in turn by the gross inadequacy of wages for the labouring classes.

During 1851 Wakley had a physical collapse and retired from the House of Commons. By 1861, he had tuberculosis. The next year, while convalescing in Madeira, he coughed up a lot of blood and died.

S. Squire Sprigge sums up Wakley's medical reforms:

In 1823, when the "Lancet" was founded, there was no Medical Act either protecting the public or regulating the medical profession; nepotism was the one prevailing force at the metropolitan hospitals; favouritism determined all official appointments and elections; the horrible trade of the resurrectionist was thriving; and the provisions for medical education were disgraceful. Within forty years the hospitals of London had resolved that ... their staffs must have merit; the Anatomy Act had abolished the resurrectionist; the Medical Act had met many of the crying grievances of the profession; and the London medical student was receiving a magnificent education. To obtain this education he had no longer to pay exorbitant fees; and to become in turn a teacher and a hospital official himself he had to buy out no predecessor. In the fight for all these reforms, Wakley led the way.

(GB)

Dr Wakley, a day to day demon on the business-world dancefloor when we speak of high-level politics, medical practise, magazine producing and being a coroner!

Chapter 4

MEDICINE AND THE ARTS

SOMERSET MAUGHAM'S DARK SECRET

When William Somerset Maugham died in 1965, he was the most widely read author of his day, a qualified doctor with no practice, but about 24 published novels, 24 plays and over 100 short stories to his name.

Robert Maugham was solicitor to the British Embassy in Paris when his fourth son, William Somerset, was born in 1874. Robert was short and ugly, while Maugham's mother, Edith, was lovely. Their nicknames? Beauty and the Beast.

The young Maugham told wonderful stories and made up games to play with friends. His mother died of tuberculosis when he was still a boy; he kept her photograph near his bed for the rest of his life. Two years later, his father died of grief and cancer. Now the family was poor. Maugham's brothers were all in England. What was to become of 10-year-old William?

His mother's maid took him to England, but Maugham knew little English. Later he described himself standing on the quay at Dover vainly calling out "Porteur. Cabriolet!"

Maugham was wished onto his childless aunt and joyless uncle who lived in the gloomy port of Whitstable in Kent. At once they sent away the maid, his only link with his happy childhood in Paris; he found himself a stranger in a strange country. His aunt and uncle were no happier to have him than he was to live with them. They were unused to children and thought him sullen. Actually, he was very shy; his stammer got worse in England. At the King's School in Cambridge he was an outsider, ridiculed because of his speech impediment. Once he had to go back to the end of a long queue at Victoria Station when he could not tell the ticket seller where he wanted to go.

In 1890, a chest infection, probably tuberculosis, interrupted his schooling. While convalescing on the Riviera, he enjoyed French literature and the short stories of Guy de Maupassant, which he later emulated.

On his return, he responded to family pressure and became a medical student at St Thomas's, the teaching hospital of South London. Polite society, including his Christian uncle, preferred to ignore the seamy side of life which Maugham saw as a student. In 1895, with a midwife, he attended 63 home confinements in the slums of Lambeth near the hospital. From this came his first novel, *Liza of Lambeth*, published even before he graduated. It is a simple, tragic love story, set among the Cockneys Maugham had observed.

Though he gave up medicine and turned entirely to writing, Maugham said that his medical training had been vital for his writing. Indeed, he spent his life observing and describing his fellow humans in a clinical, detached way. Though he himself remained withdrawn, Maugham did reveal himself in his works. He wrote an autobiographical novel, *Of Human Bondage*, about his unhappy childhood. The young hero of that book had a clubfoot, while Maugham himself had suffered similarly from his stammer.

By 1908, Maugham had four plays running in the West End.

During the First World War, he was nearly killed by a shell at Ypres. He worked in the ambulance service and described the injuries with which they dealt: "There are great wounds of the shoulder, the bone all shattered, running with pus, stinking ... gaping wounds in the back ... where a bullet has passed through the lungs."

Maugham met Gerald Haxton, a handsome, dashing young American. He also turned out to be a wild gambler and alcoholic, excluded as undesirable from England. Haxton became secretary and close companion to Maugham.

In 1916, Maugham and Haxton sailed to the South Seas. A Miss Thompson was on the same steamer from Honolulu to Apia. A quarantine inspection delayed them and all the passengers had to spend days cooped up together in a scruffy lodging house. The constant rain kept pounding on the corrugated iron roof. This meeting led to Maugham's famous story *Sadie Thompson*, later renamed *Rain*, which led to several film versions.

Later, in the war, Maugham spent two years on secret government missions. He went to Russia and saw the politician Kerensky. Maugham's mission? Nothing less than to prevent the Bolshevik revolution and so keep Russia in the war.

Later he drew on these experiences in his stories about the secret agent Willie Ashenden. Maugham used many people whom he met as a foundation for his writing. Indeed, fact and fiction, he said, were closely interwoven until he no longer knew which was which.

I find it striking that some biographies fail to mention Maugham's bisexuality. In his day, gays stayed in the closet. Society's attitude was: "Men who are like that shoot themselves". Maugham had been 22 when Oscar Wilde was tried in 1895. Wilde then spent two years doing hard labour in Reading Gaol and died soon after. An ambitious author like Maugham could not afford to rock the boat.

At the age of 42, Maugham married the divorcee Syrie Barnardo, daughter of Dr Barnardo. They had a daughter, to whom he was a poor father: "I have a notion that children are all the better for not being burdened with too much parental love."

Syrie could never dislodge Haxton (whom she called "a liar, a forger and a cheat") from Maugham's affections. They divorced in 1927. Much later, when Maugham was 80, they met again by chance. Syrie said their divorce had been a sad mistake. He replied that their mistake had been to marry in the first place: "The tragedy of love isn't death or separation. The tragedy of love is indifference."

So why had Maugham married at all? "You see, I was a quarter normal and three-quarters queer, but I tried to convince myself it was the other way around. That was my greatest mistake." At his daughter's wedding, Maugham found the groom "a most beautiful young man".

In 1944, aged only 53, Haxton caught tuberculosis and drank himself to death. Maugham was heartbroken. Later he found a new lover and secretary, Alan Searle. Maugham had to bear the lifelong contempt of his older brother Frederic, who rose to become Lord Chancellor. Frederic shared the common view of gays; moreover, in polite society, writers, especially writers of fiction, had no social standing. Interestingly, Frederic's own son was Robin Maugham, who became well known as a gay author and even wrote about his uncle.

Public opinion was kinder than the critics to Maugham's reputation. Maugham himself said that in his 20s, critics had called him brutal; later in his 30s, they called him flippant, then cynical, then competent and finally superficial.

At his villa on the French Riviera, he entertained people like Winston Churchill, Max Beerbohm, Noel Coward and General Eisenhower. But as he grew older, he grew more bitter. Maugham had undignified public battles not only with his ex-wife Syrie, but also with his daughter Liza.

At 91, he told his nephew Robin, "You know, I'm at death's door. But ... I'm afraid to knock ... Dying is a very dull, dreary affair. And my advice to you is to have nothing whatever to do with it." Maugham died in 1965.

(GB)

PEPYS AND GIBBON: THE UROLOGICAL VAGARIES OF TWO FAMOUS PATIENTS

(Adapted from the Harris Oration given to the
Australasian Urological Society in 1986)

The therapeutic arsenal available to the medical practitioner of two or three hundred years ago was pretty small. There was bloodletting, often to virtual exsanguination, especially if you were a VIP; the application of leeches in unspeakable places; clysters (or enemas as we now call them) which usually contained the leaves of exotic plants, the stomach contents of a rare Eastern animal, or some other impossibly expensive or unobtainable compound; blistering; poultices; and precious little else.

Surgery did exist but was almost completely confined to the tapping of fluid from various body cavities (or paracentesis, to give it its proper name) and the much more dramatic surgical tour de force of lithotomy, or cutting for the stone. This later was a gruesome procedure used to deal with the then apparently fairly common condition of bladder stones. If the instruments used were new then not only were they sharp but comparatively sterile (well, clean) so your chances of survival were markedly better.

It must be admitted, however, that because the surgeons had so few procedures on which to hone their art, they were well practised at those which they did perform, and they operated with speed and dexterity as well as with a certain dramatic flair. If you did come through the ordeal, then the bonus was that you had a story with which to regale your friends for life. It was the coming through that was the problem.

Two famous men of letters of the seventeenth and eighteenth centuries each endured one of these two surgical procedures associated with the urinary tract, and have left written memories of the experience for us all to read and wince over. So let us for a time abandon our cosy world of gene manipulation, organ transplants and cardiac monitoring and enter a period of frontier-style and folksy medicine with no anaesthesia, no knowledge of infection and little caring finesse, and see what happened to each of these men when their different diseases were surgically treated.

Follow me, then, back past a few genito-urinary landmarks: past, for instance, the 1930s when Harry Harris was pioneering surgery on the prostate in Sydney, and getting a world reputation for himself by so doing; past 1895 when Harrison did the first vasectomy on a human; back past 1829 when Ashley Cooper did the first vasectomy of all — on a dog (I am glad that was the order of events). Let's keep going on past 1812 when Napoleon was reported to have lost momentum and paused fatally at the gates of Moscow while he got over an attack of renal colic caused by kidney stones. We can give a sideways glance at the First Fleet in 1788 under Captain Phillip while we dash through the reign of George III, a monarch who noted three times in his 80 years that his urine had turned blueish red on standing, a clue which led to the proposal that his so-called madness was a result of porphyria, a rare metabolic condition with renal manifestations. And we can keep going past Queen Anne, who at the time of her death was so swollen with dropsy from kidney disease that she had to be buried in an almost square coffin.

So into seventeenth-century England, specifically London, at last descending by passing over the Great Fire and Plague to alight gently from our time capsule in 1658, specifically on 26 March, the day on which Samuel Pepys was cut for the stone.

Samuel Pepys

Pepys was born the son of a London tailor in 1633. His mother's brother was a butcher, and other relatives were "in trade". But he overcame this despised and downmarket background to become, in his time, England's first Secretary of the Navy; Member of Parliament for Howich; graduate of Cambridge University; Master of Trinity House; Baron of the Cinque Ports; Master of the Clothier Company; President of the Royal Society; confidant of kings and at ease in any company the cream of seventeenth-century England society could produce.

In 1655 he married a 15-year-old Irish girl, Elizabeth. She was said to be untidy, profligate and quarrelsome, though Samuel thought she was a full-bosomed beauty — few were better judges of female charms than he — and "very good company when she was well". Mind you, in her husband she had a lot to put up with, and a little peevishness could perhaps be excused on the grounds of provocation. It seems that she also suffered from painful periods, or dysmenorrhoea. She died at the early age of 29 "from fever"; unspecified, but malaria was endemic in England at the time and could have been the cause.

Samuel Pepys himself, of course, is best known for his diary which he began on 1 January 1660. This, you will note, is two years after the date of his operation. The rigour of the surgical experience was never far from his mind, for, following the operation in 1658, he resolved thereafter always to celebrate its anniversary. And, sure enough, on 26 March 1660 he records, "This day it is two years since it pleased God that I was cut for the stone at Mrs Turners in Salisbury Court. And did resolve while I live to keep it a festival as I did this last year at my house, and for ever to have Mrs Turner and her company with me."

Salisbury Court, incidentally, is a short street near St Bride's Church, off Fleet Street. Pepys was born in the street, and the novelist Samuel Richardson wrote his groundbreaking book *Pamela* while living there about a hundred years later. It is now packed with office blocks with no sign of any of the old houses.

Two years to the day after this first entry, the diarist wrote, "Up early, this being by God's blessing the fourth solemn day of my cutting the stone. At noon came my good guests. I had a pretty dinner for them, viz. a brace of stewed carps, six roasted chickens, and a jowl of salmon hot for the first course; a tansy [a kind of herb] and two neat tongues and cheese second, and we're very merry all afternoon, talking and singing and piping on the flageolet. We had a man cook to dress the dinner and sent for Jane to help us." One feels he should have added "And so to bed", but he did not.

Anxiety regarding the presumed calculus, or stone, started in 1653 while he was an undergraduate at Cambridge. He had been walking with a friend on a very hot day, eventually to arrive at a well where he downed copious draughts of cold water. Soon after returning home, Pepys suffered a severe attack of renal colic. After two or three very uncomfortable days the pain eased as the stone passed into the bladder where it remained. Knowing the mortality of surgical intervention to be about 20% at the time, sufferers did not lightly undertake to have the offending stone removed in case the treatment was much worse than the complaint. So it was left well alone.

After five years of indecision with recurrent bouts of pain, enough was enough, and Pepys enlisted the aid of Thomas Hollyer to remove the irritation. Hollyer was surgeon and lithotomist to St Thomas's Hospital, then situated in Southwark on the south side of the Thames, and a man who had been said to have cut 30 persons in one year without one death. Regrettably, the next four went to meet their Maker straight from the operating table. As I say, I think the mortality was a matter of how old and contaminated the instruments were, rather than any lack of surgical dexterity. Nonetheless, to be the next in line after this must have been an unnerving experience for all concerned.

The method of doing the job is well recorded, and for the non-urological minded and for those who have finished their lunch, let me briefly describe the procedure. The patient was placed on a table with the head raised and the buttocks projecting beyond the end. The legs were flexed at the knees and held securely in that position with the aid of a rope. The resulting position was as inelegant then as it is today when used for some gynaecological operations. Indeed, it is still called the "lithotomy" position on such occasions, in memory of its original use. It comes from the Greek *lithos,* meaning a stone and *tomia*, cutting.

Two or three assistants held the patient securely as movement at a critical moment may have resulted in a castration rather than a lithotomy, which was certainly not the object of the exercise. Some sedation was then administered in the form of extract of mandrake root or solution of opium. This usually proved to be so inefficient that it was administered more to stiffen the surgeon's resolve than to mollify the victim.

The actual operation involved either inserting the finger into the rectum to steady the stone, or pulling it down to bulge at the skin area in front on the anus, the so called "perineum". This skin was then incised and the stone either flicked out or seized with forceps.

There was an alternative and more refined method in which a grooved staff was passed down the passage from the bladder to the outside, or urethra, until it reached the neck of the bladder. When the staff was felt in the perineum the surgeon cut down onto it, entering his knife in the groove and carrying it up to its end in the bladder entrance. The stone was then located and pulled.

The skill lay in the rapidity with which the operator could locate the calculus and get it out; in practised hands it took less than a minute. But whatever method was used and however quick, the whole affair must have been terrifying and gruesome, and well worth celebrating each year with salmon jowls, stewed carp and the like.

The stone removed from our hero was two ounces (about 56 grams) by weight and after five years rolling around in the bladder had reached the size of a tennis ball. That, of course, refers to the tennis ball used in "real tennis", a lesser size than that used in the "lawn" tennis today, you will be relieved to know. All this is public knowledge because a fellow diarist, John Evelyn, saw it when he took Pepys along to cheer up his brother who was also affected by a stone and understandably hesitant over surgery. Pepys showed the sufferer the offending pathological specimen and encouraged him to go through with the operation.

The concretion itself was kept for years by the smug owner as a kind of talisman. Indeed, he had a special case made for the object, the better for its display. It seems he showed it off as a nineteenth-century Heidelberg student would a duelling scar — with insolent pride.

Despite the visible reward of the surgical virtuosity (though truth to tell most lithotomists kept a spare stone about their person to cheer up wrung-out patients in case of failure), Mr Pepys continued to suffer from renal colic for the rest of his life. Not unreasonably, he did not feel up to another dash on the operating table, and he tried to ward off attacks by the much more harmless expedient of keeping a hare's foot in his pocket. Success from this was equivocal until it was pointed out to Samuel that the charm was not working efficiently because it was lacking a joint. When this was rectified the ever honest diarist recorded, "I no sooner handled the foot but my belly began to be loose and to break wind, and whereas I was in some pain yesterday, and in fear of more today, I became very well and so continued." He seems to have got his vital systems mixed up.

When Pepys eventually died aged 70 in 1703, his left kidney was found to be disorganised and contained several calculi linked together and weighing a total of four ounces (about 113 grams).

Samuel Pepys never sired any children, despite his well-recorded sexual athleticism. It is thought that during the operation the tubes which carry sperm from the testes, or vasa deferentia, were injured, thus rendering him sterile, but certainly not impotent.

When he crossed his legs for any length of time he got a mild inflammation in the upper part of the tubular system of the testes, a condition called epididymitis. This caused some discomfort and was a source of innocent merriment to those of his friends in on the secret. He described it with less than politeness in the diary. A cold snap in the weather would also bring on perineal pain. It has also been claimed that his more or less constant state of sexual excitation was due to the continual irritation of the perineal scar. So when all else fails, have a lithotomy.

There are still in existence two prescriptions given to the distinguished patient after the operation. The main constituents were lemon juice and syrup of radish; rather insipid additives by the rip-roaring Restoration standards of the day, I would have thought.

In 1662 Pepys dined at the Chirurgeon's Hall in London and saw for himself the anatomy of the genito-urinary tract laid bare in a dissected and preserved specimen from, as he says, "a lusty fellow, a seaman who was hanged for robbery. I did not touch the body, but me thoughts it was a very unpleasant sight."

He stopped writing his diary in May 1669 because he thought the concentration was affecting his vision. It was put down in a kind of cipher and the code was not cracked until a translation appeared in the 1820s. In the 1 250 000 words of his diaries Pepys recorded great social events as well as details of his day-to-day living. He never seemed to suffer a dull moment and this curious, hardworking, pleasure-seeking and lithotomised man produced not just diaries but supreme works of literary art.

Edward Gibbon

Exactly the same can be said of the second famous patient, for he wrote what is, I suppose, the definitive book on history in the English language; the yardstick by which similar works are measured.

So come forward with me a hundred or so years from the time of Pepys and into the eighteenth century, a period redolent with the names of many great pedagogues — Newton, Defoe, Swift, Goldsmith, Johnson, Boswell, Pope, Reynolds, Handel. Follow this line and towards the end of the century you arrive in the presence of another massive intellect, Edward Gibbon, author of the *Decline and Fall of the Roman Empire*, England's greatest historian and our second patient.

His medical history is more quickly told, for, unlike Pepys, Gibbon was a self-effacing and self-conscious man. Indeed, for

32 years up to 1793, the year before he died, he deluded himself into thinking that he had kept secret a physical condition which, with the skin-tight trousers and cut-away coats then fashionable, was in fact obvious for all to see. For Gibbon had a hydrocoele, a pathological condition of the scrotum which produces a fluid accumulation in that organ. The point in selecting this particular hydrocoele for mention is not because of its intimate association with a famous sufferer, but the fact that it was not just any old hydrocoele, but one of such monstrous proportions that it hung to its distinguished custodian's knees, and has been freely canvassed as the biggest and best on record — a kind of Bradman among hydrocoeles.

Edward Gibbon was born in Putney, then a village outside London, in 1737. His father was a man of means with the extravagant habits such wealth can bring. Of the seven children born to the Gibbons, Edward was the only one to survive to adulthood, this despite the fact that he was a weakly child and easily bullied by his schoolmates at Dr Wooddeson's institution at Kingston. Later he went to Westminster School, but his attendance was fitful, as he spent much time either recovering from one illness or coming down with another. When he was 16 his health suddenly improved, his prostrating headaches vanished and he went up to Oxford.

He never married, although at one time he did fancy a Miss Suzanne Curchod, but she dallied so long over making any firm commitment his passion faded and the moment passed. In any case, his father opposed the union, and Edward, perhaps with some relief, later wrote, "I sighed as a lover, but obeyed as a son."

If he had married, the story of the hydrocoele would probably never have occurred, for any self-respecting partner would have hurried him off to the doctor very early in the piece. In the event its record-breaking presence of over 30 years' standing was ultimately revealed to Lord Sheffield, a close friend of Gibbon, in a beautifully circumlocutory letter from the overburdened wretch. It reads in part, "Have you ever observed, through my inexpressibles [the word trousers was quite taboo] a large prominency circa genitalia, which, as it was not at all painful and very little troublesome, I have strangely neglected for many years?" What grace and finesse! What sensitivity and how humble! No-one's aesthetic sensibilities could possibly be injured.

Sheffield, on the other hand, was much more downright and forthright. A Dr Farquhar was summoned who in turn called out Mr Cline, the surgeon. The mass was viewed, prodded, walked round and wondered at. It was pronounced a hydrocoele and drainage was recommended before a barrow became necessary to support it about town.

So 4.5 litres or eight pints was withdrawn by the surgeon on 14 November 1793. This is about the capacity of the largest cardboard wine cask. The swelling was diminished merely by half, but what remained formed a soft irregular mass, which puzzled the attendants. Evidently, there was more there than the fluid.

Two weeks later a second tapping was assayed, and although this effort was more probing and more painful it produced only 3.4 litres. This is not the stuff of which records are made, so more rest was ordered to see what would happen.

Events took a turn for the worse, for a few days later the area became painful and movement difficult. Inflammation appeared and Gibbon became feverish. On 13 January 1794, although the mass was now ulcerated, a third tapping was undertaken and 6.8 litres or 12 pints withdrawn, making a grand total of 14.7 litres. That's more like it. But regrettably it was all too much, and on 15 January pain returned. The following day the peerless Edward Gibbon suddenly collapsed and died.

So it appears that he originally had two conditions, the fluid accumulation and the softish mass of a hernia which was undetectable behind all that liquid. As he lived a quiet life, was of a placid nature, took no exercise and had an unruffled mind, he was never incommoded by the growing monstrosity. The tapping had almost certainly introduced infection and the distinguished historian died of complications from this, namely peritonitis and septicaemia, and not the hydrocoele per se.

A postmortem was done by Cline. He noted that the tumour extended from groin to knee. The upper part was occupied by globules of fat normally fixed to the bowel (the omentum), as well as the greater part of the large bowel itself. In parts these areas were gangrenous. The stomach, usually located under the ribs in the upper abdomen, was in the groin at the entrance to the hernial sac. Cline rather laconically concluded his report by stating that the other viscera seemed to be in a perfectly sound condition. In truth there was precious little left to display pathology; I wonder what he wanted for his money. In passing, Henry Cline was later to be tutor of that other distinguished man of letters, John Keats, while the poet was a medical student at St Thomas's Hospital, London.

Edward Gibbon was a cultivated man who wrote in the grand style of the century in which he lived and withal affected an air of learned and untroubled candour. When the first volume (of six) of his magnus opus appeared, Horace Walpole, despite a prostrating attack of gout he was enduring at the time, was moved to observe, "Lo, there has just appeared a truly classic work ... The style is as smooth as a Flemish picture ... "

The opening paragraph is compelling and immediately awakens interest, and in part reads "... the Empire of Rome comprehended the fairest part of the earth, and the most civilised portion of mankind. The frontiers ... were guarded by ancient renown and disciplined valour ... Their peaceful inhabitants enjoyed and abused the advantage of wealth and luxury. The image of a free constitution was preserved with decent reverence."

He closed his life's work when he wrote in 1787 words which proved to be prophetic:

The present is a fleeting moment, the past is no more; and our prospect of futurity is dark and doubtful. This day may possibly be my last; but the laws of probability, so true in general, so fallacious in particular, still allow about fifteen years [in the end it turned out to be seven]. I shall soon enter into a period which, as the most agreeable of his long life, was selected by

*the judgement and experience of the sage Fontenelle ... in
which our passions are supposed to be calmed, our duties
fulfilled, our fame and fortune established on a solid basis ...
and this autumnal felicity might be exemplified in the lives of
Voltaire, Hume, and many other men of letters. I am far more
inclined to embrace than to dispute this comfortable doctrine.
I will not suppose any premature decay of the mind or body;
but I must reluctantly observe that the two causes, the
abbreviation of time, and the failure of hope, will always tinge
with a browner shade the evening of life.*

He was 57 when he died.

Let me debrief you now, and rapidly bring you back into the
twenty-first century. On its return, our time machine has been
programmed to set you down at the entrance to Tower Hill
underground station. Let the family go and visit the Tower
standing by the river in front of you. For your part, turn and walk
towards Fenchurch Street, into Seething Lane and then Hart Street.
There, in a small well-manicured garden, you will see a plaque set
beneath a tree. It simply states that Samuel Pepys worked and lived
on this site three hundred or so years ago. The site is now a green
oasis in the bustling heart of the City of London.

If you turn and look diagonally across the road you will see the
church of St Olaves, tucked away between towering office
buildings. In it are buried Samuel and Elizabeth Pepys and busts of
their likenesses can be seen there. The skulls on the gate of the
church signify that it was used as a burial ground during the Great
Plague in 1665, a fact recorded by Pepys. As the Great Fire was
blown away from the building, the church was spared, as was
Pepys's house and the Navy Office. They were razed in another
conflagration about 20 years later when 30 houses, including theirs,
were destroyed.

Leaving the family still queuing to look at such hackneyed
attractions as the Crown Jewels, Henry VIII's armour and the tame
ravens, be more discriminating and go west to St James Street near

Piccadilly in the heart of London's clubland. Seek out number 76, for it was here that poor Edward Gibbon died. Incidentally, to add to the lustre of the address, a few years later here too lived Lord Byron.

Having done all that, and as you go back to rejoin the rest of the world, you can look back and reflect on two people who actually existed, and who laughed and loved and were curious and bore grudges, but who, perhaps above all, despite their pre-eminence in contemporary society, were subject to urological ailments, insults and vagaries just like you and me.

(JL)

What killed Mozart?

It was with the appearance of Peter Shaffer's play, Amadeus, *in 1979 that the once widely held supposition that Mozart had died of poisoning again came to public notice.*

M. KEYNES, "THE PERSONALITY AND ILLNESSES OF WOLFGANG AMADEUS MOZART", *JOURNAL OF MEDICAL BIOGRAPHY*

It may prove difficult to dissuade the public from the current Shafferian view of the composer as a divinely gifted drunken lout, pursued by a vengeful Salieri.

H. C. ROBBINS LANDON, *1791: MOZART'S LAST YEAR*

Despite the spilling of gallons of ink, doctors still disagree about Mozart's various illnesses and especially about the cause of his death.

Wolfgang Mozart was born in Salzburg on a freezing January day in 1756. Of the seven children born to this family, only two survived to adulthood. His father, Leopold, was a scholar, distinguished violinist and composer who also directed the orchestra of the Archbishop of Salzburg.

Mozart began to play the harpsichord at three; at five, he had mastered the violin, composed an andante and allegro and given his

**Even from an early age,
Mozart impressed all around him.
He's seen here taking his Kindy class
in an entirely new direction!**

first public performance. At six he played at the imperial court of the Empress Maria Theresa.

Leopold paraded Mozart and his sister Maria Anna on an extended tour of European courts. A nobleman wrote: "We fall into utter amazement on seeing a boy of six at the keyboard and hear him play ... like a grown man, and improvise moreover for hours on end out of his own head ... I saw them cover the keyboard with a handkerchief; and he plays just as well ..."

Soon Mozart was in London playing for King George III and meeting J. C. Bach. He performed at court, in public and in churches. At 13, he toured Italy, where Pope Clement XIV made him a Knight of the Golden Spur. As a child prodigy, Mozart lapped up attention. But there was a downside: travel by draughty unheated carriages over rough roads with overnight stops at dodgy inns. Some believe that these hardships affected his health.

As he grew up, the establishment did not respond to him as warmly as before. Mozart quarrelled with his employer, the Prince-Archbishop of Salzburg. Finally, in 1781, the archbishop sacked him. For most of his remaining ten years, Mozart had neither a regular post nor a regular income. "To the end of his life he was convinced that he would find a good and profitable post ... to the end of his life he was wrong." (McLeish & McLeish, 1978)

How much longer, happier and even more productive might Mozart's life have been if he had enjoyed a regular income?

Basing himself in Vienna, he became a freelance artist. He earned enough to live, but lived beyond his means. To make ends meet, he borrowed money, sold his compositions, played piano and violin at many concerts, taught piano, and arranged dance music and salon pieces. In 1782, he married Constanze Weber; his loving but domineering father, Leopold, disapproved of this choice. Only two of their six children survived infancy.

Mozart had many friends and liked puns and practical jokes. But some people called him arrogant, capricious, tactless and spiteful. Was this because he had a chip on his shoulder when dealing with the privileged classes, which valued birth above genius?

What of Mozart's health? At six, he had some kind of rheumatic fever or scarlet fever. At least twice he also had feverish illnesses when his feet and knees were so painful that he could not walk. Once he nearly died of typhoid. When aged 12, he had an eruptive fever, diagnosed as smallpox, which left him with pock marks on his face. Later he had a recurrent sore throat and pain that may have come from his kidneys. In 1784, he had rheumatic fever for six weeks.

Some writers state that ill health and overwork sometimes made him depressed. Others say he must have been healthy, since he managed to compose almost 300 works between 1780 and 1790. But in late 1791, he fell sick with weakness, pallor, pain in the loins, depression and paranoid delusions. On 20 November, he told Constanze that the famous "Requiem" (which he never finished) was for himself and that he was being poisoned. But there is little to suggest that Salieri (or anyone else) poisoned Mozart. The repeated bleedings and vomiting induced by his doctors may have hastened his end.

Over his last 15 days, he had fever, painful swelling of the hands and feet and then vomiting, diarrhoea, and what he called the taste of death on his tongue. Constanze crawled into his bed, hoping in vain to catch his illness and join him in death. Even the four doctors who signed his death certificate could not agree on the cause of death and there was no postmortem. Nowadays one popular diagnosis is recurrent rheumatic fever with heart failure. But there have been over a hundred others which include blood poisoning, severe kidney failure, typhoid, typhus, meningitis, an overactive thyroid gland, syphilis, and poisoning with mercury or antimony.

About his musical genius there is no dispute; he excelled at every medium of his time. Many critics honour Mozart as the most universal composer in the history of Western music. Though he died before turning 36, he left us over 600 works. Many regard *Don Giovanni* as the world's greatest opera. His seven piano concertos make him a pioneer of this form.

But no mourners went with his funeral wagon to the cemetery where he was buried in an unmarked common grave. Visitors today can still see a skull, said to belong to Mozart, at the Salzburg Mozarteum.

(GB)

ROBERT LOUIS STEVENSON AND HIS CHEST

Robert Louis Stevenson was an adventure storyteller, South Sea traveller and chronic invalid. A consummate writer, yet an odd man, who, while doggedly pursuing relief from his presumed tuberculosis, at the same time appears to have seen his invalidism as a merciful preparation for death. He once wrote, "I have wakened sick and gone to bed weary. I have written in bed, written out of it, written in haemorrhages, written torn by coughing. My case is a sport. I may die tonight or live till sixty."

He actually died in 1894 at the age of 44 and not, in the end, from tuberculosis, but from a sub-arachnoid haemorrhage or haemorrhage onto the surface of the brain.

Stevenson was born in Edinburgh in 1850, the only son of a lighthouse engineer and tubercular mother. He was a frail youth whose precarious physical state made regular schooling difficult. Nonetheless, he went on to Edinburgh University to study engineering, but then transferred to law, becoming an advocate in 1875. He never practised his profession, but pursued his true inclination — writing.

A wandering lifestyle in search of a healthy climate followed. At first, any benefit gained was offset by his impoverished lifestyle and the writing he carried out was punctuated by bouts of pneumonia, pleurisy and coughing up blood. His roamings took him mainly to France, where, among other pieces, he wrote *Travels with a Donkey*. In 1876, while still in France, Stevenson met an American divorcee, Fanny Osbourne, whom he followed to America and eventually married in 1880. The pair travelled extensively together, and the masterpiece which at last brought him fame in 1883, *Treasure Island*, was written while oscillating between Scotland and Switzerland.

Struggling with his ailments, he and Fanny doggedly chased the sun and warmer climes, eventually deciding to settle in the South of France. They could not seem to take a trick, though, as insult was added to injury when a cholera epidemic drove them out. Bournemouth in the South of England was the nearest they could find to an equitable and harmonious existence. Paradoxically, it was during the three years spent there (1884–86) that Stevenson was both at his most consumptive and literally productive. So marked were his chest symptoms that his doctor forebade the author to speak aloud for fear of precipitating a bleed. I suppose at least this allowed him to get on uninterrupted with *Kidnapped* and *The Strange Case of Dr Jekyll and Mr Hyde* which emerged from that period.

Jekyll and Hyde was written in three days of frenetic activity. The author destroyed the first draft, but immediately rewrote it and corrected another 30 000 words written, of course, in longhand in three more days and nights. It seems that for him the fire which wasted the body made the mind shine with brighter light. Or was it something else; his treatment, perhaps?

Even the bacterial cause of tuberculosis, let alone specific therapies, were unknown at that time. Opium, or its alcoholic tincture, laudanum, was commonly given for cough suppression. This certainly gave symptomatic relief, but significantly in the kind of author Stevenson was, it also gives a relaxed mind, and allows flights of fancy. We know for certain that Stevenson took morphine, for in 1884 he wrote, "...the morphine I have been taking ... moderates the bray, but I think sews up the donkey".

But there is more to it than that. In the July 1885 edition of the highly regarded medical journal the *Lancet*, the newly researched drug cocaine was given an enthusiastic write-up, especially as a cure for hay fever and asthma. In September it got a further warm appraisal as an application to the larynx for chronic cough.

Now we know from her son, Lloyd Osbourne, that Fanny Stevenson was an avid reader of the *Lancet*. Together with Robert's doctor, Thomas Bodley Scott, she could well have persuaded the invalid to take the new drug as a therapeutic trial for his chronic cough. Little was known at the time of its side effects, but perhaps significantly two months after the *Lancet* article appeared, during September 1885, the chilling story of *Jekyll and Hyde* was written. As we have seen, this was during a period of unsurpassed psychic energy, and quite possibly the author's drug intake may have been at the bottom of it all. It is also worth noting that in the story Jekyll takes a powder to transform his personality: cocaine came in that form.

As is well known, Sherlock Holmes, Conan Doyle's fictional master sleuth who appeared on the bookstalls at a slightly later date than Stevenson's work, took cocaine throughout much of his illustrious career. Conan Doyle was a general practitioner in

Southsea near Portsmouth and presumably knew all about the new drug which was quite the rage in the 1880s and 1890s.

The odd thing is that normally Stevenson was a brilliant and sensitive writer of adventure stories, whereas *Jekyll and Hyde* uncharacteristically belongs to a different and more brutal genre. Was it written under a different driving force? Who knows?

Eventually, Robert's ill health forced the Stevensons to leave Bournemouth. In 1887 he and his wife went to America and in 1888 to the South Seas, wandering from place to place before finally settling in Valmia, Samoa.

The balmy climate suited the author and on 3 December 1894 he was working on *The Weir of Hermiston*, a book which some contend may have become his masterpiece. That night Stevenson got up from the dinner table, cried, "What's that?" and clutched his head. With a sub-arachnoid haemorrhage a small blood vessel situated at the base of the brain ruptures and often the sufferer describes a sudden crashing noise in the head. I am sure that was what the writer was experiencing. The condition is commonly fatal and in fact Stevenson died within the hour from this cerebral bleed.

There is one oddity about the great man's condition, which troubled Dr Livingstone Trudeau, a tuberculosis expert with whom Stevenson lived for a short time in New York State. If the writer's consumptive condition was as putrescent as history would have us believe, how was it that he lasted for as long as he did?

Mind you, as has been observed, the genesis of tuberculosis was ill understood in 1893, but at least the existence of the tubercle bacillus was known and could be recognised under the microscope. Its identity had been first revealed to the medical world on 24 March 1882 when Robert Koch, the German bacteriologist and discoverer of the *Mycobacterium tuberculosis*, read a paper giving its details at the Berlin Physiological Society. Trudeau was aware of this and his scepticism about the severity of the writer's malady was fuelled by the fact that he never found any of the diagnostic so-called "acid fast tubercle bacilli" in Stevenson's spit. If his observations can be relied upon, it seems a pretty strong argument against that disease being the cause of his chronic ill health.

Perhaps after all Robert Louis Stevenson was "a sport", a one-off, as he used to claim. Or could there have been another cause for his persistent signs and symptoms, possibly bronchiectasis, a chronic low-grade and debilitating lung infection caused by a germ other than tuberculosis? We shall never know, but maybe in the end, all that enervating travelling was unnecessary.

(JL)

THE MACHO LIFE AND DEATH OF ERNEST HEMINGWAY

I still remember how thrilled I was as a teenager to read Ernest Hemingway's short novel *The Old Man and the Sea.* But Ernest's own life was as remarkable as his fiction; indeed, fact and fiction merge together. For instance, he boasted of sleeping with Mata Hari, whom he never even met. Anthony Burgess wrote: "Hemingway the man was as much a creation as his books."

Before he turned five, Ernest told his grandfather that he had singlehandedly stopped a runaway horse. The old man predicted that with such an imagination, the boy would become either a jailbird or famous. His father taught him to love the outdoors; Ernest got his first shotgun when he was ten. At school, he enjoyed football and boxing; the latter remained a lifelong habit. But at training he twice received a broken nose and an eye injury left him with a slight squint. Twice he ran away from both home and school.

When the First World War broke out, he tried to enlist. When his eye injury debarred him from the army itself, he volunteered to drive an ambulance with the Italian army. He was in the Italian frontlines in 1918 when a shell landed only a metre away and riddled his right leg with shrapnel. Ernest, still only aged 18, carried a wounded soldier back to the first-aid dugout.

The next three months he spent in a military hospital, where
he won out against his doctors who wanted to saw off his leg.
By his bedside he kept the 227 metal fragments they had
removed. Ernest gave these to visitors as souvenirs. The two
medals he had won he threw into the bowl "with the other
scrap metal".

Ernest returned to America as a hero but suffered nightmares and insomnia and was already drinking. During the 1920s, he suffered malaria, anthrax, a broken arm and an injury to his good eye. Aged 22, he married 30-year-old pianist Hadley Richardson. She admired his "boxing, fishing, writing ... getting war medals ... charm, good looks".

Ernest roamed Europe as a roving reporter for the *Toronto Star* and interviewed Mussolini. In Spain he began his lifelong obsession with bullfighting and bullfighters. In 1924, he risked life and limb in the running of the bulls through the streets of Pamplona. Later he used bullfighting as the background for *Death in the Afternoon.*

By 1926, Ernest was having an affair with an American reporter, Pauline Pfeiffer, who became his second wife. In 1933, they spent five months on safari in Africa, where he bagged four lions, a buffalo and a rhinoceros. Back home, he caught a huge shark, but shot himself in the foot while gaffing it.

In 1936, the Spanish Civil War broke out. Having seen Fascism in Italy, Ernest opposed Franco and sent $40 000 to the Spanish Republicans for medicines and ambulances. As a non-combatant war correspondent, he stayed at the Hotel Florida in Madrid. When a shell hit the hotwater tank at night, guests rushed out of their rooms. Among them were Ernest and Martha Gellhorn; Martha later became his third wife. But Pauline still wanted to save her marriage and threatened to kill herself when he asked for a divorce.

By 1938, his drinking had affected his liver. He achieved a hit with *For Whom the Bell Toll*s, a novel based on the Spanish Civil War.

When the USA entered the Second World War, Ernest wangled approval to arm his large deep-sea fishing boat with machine guns and chase German submarines. The crew carried grenades to throw down the conning towers of submarines; whether they even saw any subs is not clear. Either way, they carried enough booze to keep them happy. Soon Martha was complaining that they drank too much and washed too little.

In 1944, he suffered concussion in London during the blackout when his car crashed into a water tank, but he slipped out of hospital and got a ride on a reconnaissance flight over enemy lines in France. Next he embarked with the American Fourth Infantry Division to cover the invasion of Normandy. The stories he sent to *Collier's* magazine were "wildly inaccurate but full of life". Though officially a non-combatant, he organised his own partisan group and defied the Geneva Convention by lobbing grenades into a cellar where Germans may have been hiding. Ernest's irregulars got 60 miles ahead of the allied army, reached Paris before the Free French troops and then skirmished with the Germans. A military enquiry failed to convict him; indeed, Ernest won the Bronze Star.

After his disastrous third marriage (to Martha) ended, he pined again for Pauline ("the best wife a man could ever hope to have"). I doubt whether even one of his four wives would have thus praised Ernest. Soon he fell in love with journalist Mary Welsh. When her husband opposed a divorce, Ernest emptied a machine pistol at his portrait.

By 1947, he had high blood pressure. On a duck hunt in Italy, cartridge wadding hit him in the eye. Doctors saved the eye with penicillin, but he lost part of his vision.

When reviewers panned his novel *Across the River and into the Trees*, Ernest threatened to crack their empty heads. He quarrelled with many of his fellow writers. Rather than discuss writing, he preferred to box or arm-wrestle them. He was also, to many minds, prejudiced against Jews.

Next he was clawed while playing with a lion!

In 1953, Ernest and Mary were lucky to survive a safari in Kenya. Depressed by Pauline's death, he was still drinking heavily. They boarded a plane to see Victoria Falls, but a flock of ibises crossed their path; the pilot turned to avoid the birds, hit an old telegraph line and crashed. Sharing one bottle of water, one of Scotch and four beers, the trio spent the night in the open near a herd of elephants. The noise of the falls drowned out the sound of the search plane. The circling pilot saw the wreckage, but no signs of survivors; reports of Ernest's death flashed around the world. But a passing riverboat picked them up. When they boarded another small plane (perhaps overloaded) this crashed and burst into flames. He butted the jammed door open with his head and injured shoulder. It was reported that Ernest ended up with a dislocated right shoulder, a fractured skull, a ruptured liver, spleen and right kidney, and first-degree burns. If all this were true, he could not have survived. Later he joked that his cracked vertebrae left him with a permanent erection.

In 1954, Ernest won the Nobel Prize for Literature. But by now, he had high blood pressure, high cholesterol, alcoholism, liver damage and kidney problems as well as failing eyesight, diabetes and impotence. An unrealistic doctor told him to cut out alcohol.

When a crew were shooting a film of the *Old Man and the Sea*, he insisted that they film a real duel with the biggest swordfish ever caught. They had to go to the Pacific and the film finally cost $60 million.

Mary threw him a huge party for his 60th birthday. Ernest had a visiting bullfighter stand still with a lighted cigarette in his mouth. Then Ernest shot it out seven times with a rifle, the butt in the man's mouth getting shorter each time.

In 1960, Ernest slid into depression. He avoided his friends and would not even go to bullfights. "He had developed delusions of persecution and was frightened, lonely, guilty, full of remorse and drinking heavily." Mary said: "He was ... exactly the opposite of what he had been before — outgoing and exuberant and articulate and full of life ... "

By now Ernest and Mary were threatening each other with guns. Four times he seriously attempted suicide before entering the Mayo Clinic for twice-weekly bouts of electroconvulsive (ECT or shock) therapy. Twice he entered the clinic, twice he talked his doctors into believing he was better and letting him go, even though Mary knew how depressed he really was.

Ernest was asked to write just one sentence for a presentation volume to the new President John Kennedy. Despite a heartbreaking whole day's struggle, he could no longer write. He blamed this writer's block on the ECT.

On 2 July 1961, while Mary was still sleeping, he went down to the basement, got out his favourite shotgun, came back up into the hallway (where Mary was sure to find him), put the double barrel into his mouth and blew out his brains. Did his long-suffering fourth wife Mary have a hand in this tragedy? Was she ambivalent about preventing his suicide? She locked all his guns in the basement but left the basement keys in the kitchen!

All his life Ernest had suffered many near-fatal accidents and injuries. Is it fanciful to see these as self-destructive? His life combined genius, the unceasing need to prove his masculinity and a lust for life, but also an obsession with blood, weapons, killing and suicide.

It is said that his mother had wanted a girl; she called him Ernestine and dressed him in girl's clothes. Fellow writer Sydney Franklin wrote: "Ernest's big problem all his life, I've always thought, was he was always worried about his Picha (penis) ... the size of a 30/30 shell."

His reputation as a writer fluctuated wildly; he was praised in the 1920s but flopped in the 1930s. It rose again in 1940 with *For Whom the Bell Tolls*; another triumph was the Pulitzer Prize for *The Old Man and the Sea*. His terse, bare prose style was probably the most widely imitated (and most widely satirised) of any writer of English in his century. Meyers (1985) sums up Ernest the writer: "He created unsurpassed images of Italy, France, Spain and Africa ... He wrote as naturally as a hawk flies and as clearly as a lake reflects."

(GB)

A VERY CLUBBABLE MAN: A MEDICAL HISTORY OF SAMUEL JOHNSON

Boswell's "The Life of Samuel Johnson L.L.D." came out over two hundred years ago in May 1791, but despite its well-recognised scholarship and personal details it is mainly from other sources that we learn about the lifelong and burdensome ill health which oppressed the great lexicographer himself. Of his many pathologies, certainly the most interesting was the one which produced in him the involuntary jerky movements, uncontrolled vocalisation and compulsive actions which caused such consternation in his friends and discomposure in his foes. It is now thought to have been due to Gilles de la Tourette syndrome. But don't let us get ahead of the story, so more of that later.

Johnson was born in 1709 and defided convention from the start, for he was delivered by "a man midwife", a very unusual health professional at the time, when midwives and handy women were the customary accoucheuses.

As an infant he developed tubercular glands in the neck, or scrofula, caught, it was said, from his wet nurse. Contemporary thinking was that the one certain cure for the tuberculosis, or the King's Evil as it was called, was to be touched by the monarch. So with this in mind, in 1712, at the age of 30 months, he was taken to London to see Queen Anne. In the event, Johnson became one of the last people in Britain to be so treated as the practice was abandoned shortly afterwards. The British Museum still has the medallion he was given by the Queen for a form of therapy which was as dramatic as it was useless. On one side the medal shows a ship in full sail and on the other side the Archangel Michael slaying a dragon. Johnson wore it all his life as an abiding testimony of the regal fingerwork. Naturally, the touching had little effect and the glands were eventually drained; the scar can be seen on his death mask.

From early childhood it was obvious Samuel had poor eyesight. The contemporary treatment was to incise the left arm to allow the bad humours to escape. The wound was kept open by inserting a small foreign body, such as a pea. This treatment was as useless as the royal medallion, so our hero remained short-sighted all his life, and, amazingly for such a literary giant, had difficulty reading and never corrected his copy. It is said he often singed his wig bending too near the candle to make out the print. The portrait by Reynolds done in 1775 shows him squinting at a book held uncomfortably close to his face.

Johnson was also hard of hearing, possibly a legacy of his incised scrofulous glands. He had a special pew near the preacher and was often heard to say, with some exasperation, "Louder, my dear Sir, louder, I entreat you, or you pray in vain."

His bouts of depression lead to morbid introspection and were so enervating that he often felt he was going mad. He referred to it as "a vile melancholy I inherited from my father", and dreaded being carried off to Bedlam, the public "madhouse". He entrusted his friend and confidant, Hester Thrale, with a padlock so she could lock him up and look after him privately if it came to that. He dreaded solitude and delayed returning home at night where he knew he would brood and be tormented by nightmares.

Added to all this, for the last 40 years of his life he had chronic bronchitis and gout. So, in the end, the unedifying picture of a broken reed emerges, far removed from the usual picture of him portrayed as a bear-like, forceful, positive man, and a towering intellect taking his place among similar geniuses, including the likes of Pope, Hogarth, Garrick, Goldsmith, Reynolds, and alive at the same time as Mozart, Beethoven, Newton and Gibbon. He died of congestive cardiac failure in 1784 and packed a great deal of living into his 75 years.

All that is pretty mundane and could have happened to any sociable scribbler, so let us hurry back to the fascinating clinical aspect of his life — his tics and gesticulations. Like many others on first acquaintance, Fanny Burney was shocked at Johnson's behaviour and wrote:

Johnson's hearing problem left him out on his lonesome (especially at Church). The final straw came when he declared that kids these days have no respect for their elders ... Johnson was now officially declared old!

His mouth is almost constantly opening and shutting as if he were chewing. He has a strange method of frequently twirling his fingers and twisting his hands [skilfully depicted, incidentally, in the 1769 Reynolds portrait of Johnson]. His body is in constant agitation, see-sawing up and down; his feet are never a moment quiet ...

Alexander Pope wrote, when his friend Johnson had been rejected as a schoolmaster, "He has an infirmity of the convulsive kind ... so as to make him a sad spectacle." Boswell also mentions how Johnson held his head to one side, shook and rubbed his left knee in a persistent circular motion.

There are numerous descriptions of his repetitive movements, but they grew fewer over the years as people became used to them, and never did the movements interfere with his motor ability, or cramp his handwriting.

The muscular movements were accompanied by grunts, clucking, sighing, whistling and continual talking under his breath, saying things like, "too, too, too", or bits of the Lord's prayer or verses from the classics or repeating snippets of conversation over and over again. The symptoms, especially this echolalia, were more pronounced when he was alone or in reverie.

The sister of Joshua Reynolds records some strange habits such as always springing over a threshold as though to see how far he could stride. Or in company, while breathing hard, he would stretch out his arm while holding a full cup of tea, to the great discomfort of those nearby. He avoided the cracks in the pavement and touched every post in the street. If he missed one he would return to make good the error.

It would seem that the picture to emerge is one of repetitive involuntary tics of limbs and head accompanied by vocalisation, but unimpared intellect. Although the well-known clinical feature of obscenities, or coprolalia, does not seem to have been very evident, it looks very much as though Johnson suffered from that well-known disease eponymously known as the Gilles de la Tourette syndrome.

Some years later, in 1825, Itard described the strange case of a seven-year-old with involuntary movements and coprolalia (literally "talking faeces"). In 1885 Gilles de la Tourette described the same person who had "ticked and blasphemed" for 60 years. He was able to add eight other cases to his collection.

The syndrome usually develops between the ages of five and ten and initially the rapid involuntary movements are characteristically round the eyes and are often complex and multiple. Later they occur in the limbs and torso. Sufferers learn to convert the movements into purposeful actions so they are less conspicuous. Johnson could do this. They are more marked in stress and disappear during sleep. Repeating words or phrases, so called echolalia, is common, but uncontrollable foul language is not universal. If the desire to utter obscenities does occur at socially inappropriate times, a growing inner tension is felt and privacy may be sought to relieve the stress; the so-called and diagnostic "lavatory coprolalia".

Tourette's syndrome is not nowadays regarded as being particularly rare, but a commonly undiagnosed disorder genetically transmitted by a specific mutated gene, and which responds quite well to drugs such as Haloperidol.

Boswell thought that Johnson's habits were "of the nature of that distemper called St Vitus Dance ... the description which Sydenham gives to the disease". Joshua Reynolds thought them psychogenic, saying, "the great business of his life was to escape from himself ... considered as the disease of the mind, which nothing cured but company". Other modern writers have forwarded psychogenic theories, but it seems there is sufficient evidence to support the Tourette theory.

Gilles de la Tourette himself was born in the French village of Saint-Gervais-les-Trois-Clochers in the Poitou district in 1855. He was a gifted child and studied medicine first at Poitiers and subsequently at Paris. He became a junior hospital resident of Jean Charcot, the great Parisian physician who is regarded as the founder of modern neurology.

He went on to become a prolific writer in the field of neurology and psychiatry, as well as in art, literature and mesmerism. The young man was brash, forthright and unconventional, and in 1896 at the age of 41 was shot in the head by a paranoid patient (deservedly so, some thought!). Thereafter, his mood fluctuated between depression and hypomania and in the end he was stripped of his academic posts. He died in a mental home in 1904.

All that, of course, was long after Johnson's time, so there never could be a firm diagnosis of the eighteenth-century sage's odd behaviour.

Nonetheless, Dr Johnson was a remarkable man who, despite his multiple lesions, enjoyed life to the full in the remarkably illustrious company of his time. Between 1748 and 1759, while compiling his dictionary, he lived in Gough Square off Fleet Street.

The house is still there and can be visited. It is just around the corner from his favourite eating place and debating chamber, the Cheshire Cheese, which is also still intact and still serving typical British fare. In it is displayed the chair in which the great man is said to have sat while holding forth with his friends.

Samuel Johnson was regarded then, as he is now, with admiration and reverence, and it may be that the glittering intellectual stimulus in which he lived and revelled not only helped him to come to terms with his maladies, but provided the impetus for his greatness.

(JL)

A RAKE'S PROGRESS: A HISTORY OF JAMES BOSWELL AND HIS SOCIAL DISEASE

James Boswell was a man of his times, being at once urbane, cultivated, industrious, resilient, and the possessor of literary skills of a rare kind. But on top of all this diversity he had one further talent — he harboured an almost insatiable sexual appetite. In fact, it was of such a degree as to be regarded as remarkable even by the licentious standards of the eighteenth century in which he lived. He is, of course, better known for the biography he wrote of Dr Samuel Johnson than the details of his own life, but he was a compulsive and frank diarist and as such has left us a penetrating insight into the manners and morals of the era, more especially his own.

In the genre of this book, one feature of those somewhat meretricious and priapic days about which he wrote is worthy of a closer look, for from it emerges not only his own rumbustious story but also the history of a medical and social condition which is usually kept veiled from decent society: gonorrhoea.

Boswell was born in Edinburgh in 1740 in a gloomy alley within the shade of St Giles Cathedral on the one hand and the

Tolbooth gaol (the "Heart of Midlothian") on the other. His father was a lawyer who became a judge in 1754 and later became Lord Auchinleck. This was not a hereditary honour, so his eldest son, James, remained plain mister for the rest of his life.

His mother was described as being extremely pious and was a key figure in James's childhood. A staunch Calvinist, she instilled into the young lad stories of hellfire and the doctrine of the eternity of punishment with accompanying flames, rather than the bliss of heaven. So he was forever consumed by guilt about his subsequent sexual jousts. Fortunately for our story, he easily overcame such dark thoughts, and led a life untrammelled by any waves of high moral tone: nothing could stop him undertaking numerous illicit bedroom campaigns.

Boswell was a moody man to the point of being a manic depressive. Consequently, he frequently displayed a zest for life to be then plunged into profound melancholy. This was a characteristic which may have explained the first bouts of rather odd behaviour in his life, namely his extraordinary religious vacillations as a youth.

He was brought up, as I say, a Calvinist, but at university, where he studied logic and metaphysics and later law, he converted to Methodism. This form of faith was the opposite to his childhood teachings, but was modish at the time. However, never one to go with the crowd, this proved to be a temporary aberration, and young James changed to become, of all things, a Pythagorean vegetarian. This rigorous and deviant discipline never seemed to catch on, so he converted to Catholicism. Confessions and such like proved to be somewhat restrictive to his lusty nature, so, after a few months, Boswell embraced Deism, a hedonistic faith given over to the pleasures of sex. This was more in his line, and he was a willing devotee for five years.

In 1764 he emerged from its teachings, presumably jaded but sated, to be admitted into the Anglican Church, an institution of which his hero, that old bigot Dr Johnson, was a pillar. The two had met for the first time the year before and Boswell developed an admiration for Johnson enough to prompt him to change his church so as to be nearer his paragon's way of thinking.

Initially, his religious anxieties seem to have been generated by uncertainty regarding the afterlife. In this context he appears to have been preoccupied with death; he rarely missed a good hanging at Tyburn. There was more to it than that, though. His heavy-handed early religious background with promise of nothing less than damnation for transgressors may have been the impetus for his rejecting the mores of so-called decent society and pursuing a life of lechery and depravity. It is this hedonistic aspect of his life and the diseases that it brought that fit more into this book.

Early on in life Boswell found himself blessed, or was it burdened, with this overwhelming sexual drive. It provided him with an identity crisis, the magnitude of which was compounded, as I say, by his religious vacillations and doubts, for the feelings generated were crude, direct and urgent. He wrote of himself, "I am of a warm constitution; a complexion, as the physicians say, exceedingly amorous; I ought to be a Turk." I have never heard a physician say that. As a result of this obsession and despite his expressed fears of unwanted pregnancies, venereal disease and moral decay, he was forever plotting new sexual adventures. He wrote, "I was laying plans for having women, and yet I had a most sincere feeling of religion."

Boswell met Rousseau in 1764 and discussed that philosopher's own admitted boundless sexual energy and fantasies, hoping for some fellow feeling, and maybe a night out. He misjudged his man, for the Frenchman, though libidinous, was more into sensual mysticism, considering pleasures of the flesh to be ephemeral when compared with their spiritual virtue. This was at variance with Boswell's approach which had no time for bandying theories in a field of activity which demanded direct, basic, uncomplicated physical congress, no questions asked, no leave sought.

The upshot was that between the ages of 20 and 29 Boswell succeeded in acquiring a string of conquests, which included taking as his mistress three married women from the upper ranks of society, having liaisons with four actresses, a brief but passionate affair with Rousseau's lifelong mistress, keeping at least three of what he called "lower class women" as mistresses and producing two illegitimate children by them, and making a swift early-morning foray upon the pregnant wife of the King of Prussia's guard. The main recorded outlet for his insatiable sexual appetite was over 60 different prostitutes in London, Edinburgh, Berlin, Naples, Venice, Paris and sundry other places. Added to all that, he claimed to have laid unsuccessful siege to many other "ladies of quality". This was all before he was 30 years old.

He records that he lost his virginity at the age of 20 during his brief period of flirting with the Catholic Church. He recorded this as taking place in the Blue Periwig tavern in Southampton Street, London, with a lady called Sally Forrester. He later wrote how he looked back fondly on this introduction into "the melting and transporting rites of love", and under the seasoned tutelage of a couple of streetwise lechers, shortly after was off on what he probably thought was to be a lifelong titillating idyll.

It was to be brought to a juddering halt by his first medical problem, an attack of gonorrhoea. This was to be the first of at least 17, probably 19, doses of this venereal disease. He managed to crowd them into what must have been an extraordinary 55 years. With a record like that he must have become quite an expert on "the clap", as it was called. So let us take a diversion from the life of James Boswell itself to look at the position of this condition in history, and at some of his fellow sufferers.

Gonorrhoea is perhaps the most ubiquitous of those infections which peak in times of war, mass migration, poverty and, perhaps paradoxically, more recently, prosperity. The first scientific observations are attributed to Hippocrates, the "Father of Medicine". He lived between 460 and 355 BC and wrote, "Those suffering from tubercle and carnosities in their pipes will get well by suppuration and flow of pus". It sounds ghastly, and probably was. To "carny" is an old verb meaning to coax or wheedle, and carnosities are probably growths causing stricture of the passage from the bladder (or urethra) and which had to be teased out.

Galen (c. 130–210) a well-regarded physician from Pergamum, now in Turkey but then in Greece, was in his time medical officer to the school of gladiators so could probably be regarded as an expert on social diseases, and it was he who gave us the word "gonorrhoea", by which he meant "flow of semen" (from the Latin *gonas*, meaning semen, and *rhoia*, meaning flux). Though an influential writer, Galen was wrong in many areas of medicine and this may well be another, for the name he gave implies that he thought it was an extention of a fighter's normal physiological processes and did not appreciate that gonorrhoea was an inflammatory and pathological process in its own right.

A contemporary of Galen's, Areteaus, was the first to recommend treatment. This included poultices, sexual abstinence, wrapping the genitals in wool and that good old standby in matters sexual, cold baths. The Persian physician, Avicenna, appears to have felt there was not enough punishment in this not unpleasant regime and augmented it by recommending that in obstinate cases a flea be introduced into the end of the urethra. The therapeutic worth of this would seem to be dubious, but doubtless it would serve to concentrate the mind on past sins.

In 1161 an Act was proclaimed in London which forbade the brothel keepers of Southwark from housing "women suffering from the perilous infirmity of burning".

The word "clap" first appeared in a manuscript of 1378 written by John Ardenne, surgeon to Richard II. Prostitutes in the Paris district of "le Clapier" were quartered in "clapisses", and the word may have come from this.

The first cases of syphilis appeared in Europe at the end of the fifteenth century, and it was thought even by the great and well-respected doctor of the Renaissance period, Paracelsus (1493–1541), that gonorrhoea was an early symptom of syphilis, and not a separate disease in its own right, an error that was to persist in many minds for two and a half centuries.

Ambroise Parè (1510—1590), the so-called father of surgery, despite his renowned clinical judgment and the fact that he served with the French army on many campaigns, was fooled by the truth that concomitant syphilis and gonorrhoea was a common presentation in fighting men. He wrote in 1564 in his work *Des Chaudes Pisses et Carnosities Engendrees au Meat Urinal*, "The clap is due to three causes — sweating, starvation and infection." He recommended bleeding, purging and, again, cold baths. Local treatment comprised an emollient ointment applied on linen wrapped round a thin candle and introduced up the urethra. He also forbade the company of women or even looking at paintings of them for fear it would heat the blood. The advice seems to have had more sadistic overtones than therapeutic worth.

That great humorist, satirist and doctor, Francois Rabelais (1494–1553) tells that the son of Gargantuan caught the clap and was given drugs "to piss away his misfortune".

The confusion of the two venereal diseases being manifestations of the same malady persisted and made rational treatment difficult. Truth to tell, it mattered little as there was no rational treatment anyway and mercury was given for both. All that particular drug usually succeeded in doing was to give the sufferer a metallic taste and greatly increase salivation, and to cause the teeth to fall out and produce mental disturbance.

In 1736 Nicholas Robinson, wise before his time, wrote, "I wish the libertines of our times would take example from the numbers that daily die martyrs to the pleasures of Venus."

As all medical students know, the great British surgeon and anatomist, John Hunter (1728–1793), is said to have sustained the fallacy of the single disease when he carried out his famous auto-inoculation experiment in 1767. He rubbed "the matter of a gonorrhoea" onto two puncture marks on his own genitals. Unfortunately, he chose the inoculum from a patient suffering from both syphilis and gonorrhoea, hence caught both, thereby declaring that they were in fact one and the same disease. Such was his contemporary eminence that he held up progress in venereology for over 30 years by this one assertion.

The state of the art was thus far advanced at the time of our leading subject, for the experiment was carried out during the period that James Boswell was acquiring his 17 attacks from one of the 2000 or so brothels which provided a working environment for the 50 000 harlots said to be in London at the time.

The two conditions were finally separated by Benjamin Bell in Edinburgh in 1793. He was a more canny Scot than Hunter. He did not carry out life-threatening and messy experiments on himself. Certainly not: he inoculated medical students instead. Their names are lost to posterity, but it seems a pernicious way of ensuring examination success.

In the nineteenth century, for the first time, hospital beds were set aside for suffers of venereal disease. In England they were known as lock wards, not because the patients had to be locked in, but because "lock" is derived from *logue,* meaning a rag such as lepers found in baskets at the gates of lazar houses and used to wipe their sores before entering. By the fifteenth century leprosy had declined to such a degree that the lazar houses provided by the monasteries for the use of lepers were half-empty and so were pressed into service for venereal diseases sufferers instead.

One of the great venereologists of the nineteenth century, the Frenchman Philip Ricord, was said to be have been an engaging personality, frank but discreet. His extroverted nature made him a sought-after lecturer, and he was famous for his "recipe for getting the clap". Politically incorrect though it may be when viewed nowadays, in part it ran:

Take a lymphatic, pale and preferably blond women. Dine with her; begin with oysters, continue with asparagus, drink heavily with white wine and champagne. You will be well on your way then. To expedite matters, dance together until you feel hot. Once the night has come on, set to work energetically; two or three connections are by no means too much — the more the better. If you live up to this programme and do not get ill, you must be under the special protection of a god.

Naval personnel have a reputation for having a girl in every port and there is no doubt that in the same century the medical chest on Nelson's HMS *Victory* contained more metal bougies for dilating the gonorrhoea-constricted urethras of the naval personnel than it had guns for engaging the enemy.

In 1879 Albert Neisser, working in the Breslau Skin Diseases Clinic, gave the first detailed description of the gonococcus, the germ that causes the complaint, and in medical circles ever since it has been known as the Neisseria gonorrhoea in his honour.

In this century antibiotics have been very successful in the treatment of the condition, but with growing resistance their days may be numbered. For a variety of social reasons, gonorrhoea is as common now, if not commoner, than in the days of that long-suffering and dedicated rake, James Boswell. So let us return to him.

We have seen his first infection was from a sexual joust with Sally Forester; the second was from Ann Lewis, a Covent Garden actress also known as Louisa, whom he had hoped would give him "a winter's safe copulation". It was not to be.

Of what symptoms and signs would Boswell have complained? Well, he would probably have admitted to a sexual contact and not have cited a variety of bizarre postulates which seem to be common among some present-day sufferers; reasons such as unaccustomed exercise, a blow from a cricket ball, strange lavatory seats or urinating on hot stones are still surprisingly common — anything, in fact, except sexual intercourse.

Following an incubation period of two to five days Boswell would have noticed some discomfort on urinating, to be succeeded by a urethral discharge, usually purulent. Indeed, Boswell's description itself cannot be bettered when he wrote, "I felt a little heat in the members of my body sacred to Cupid, and by the next day it too, too plain was Signor Gonorrhoea again." There would also have been frequency of passing water which was probably due to the infection, but could have been due to guilt, remorse or introspection.

As we have seen, Boswell did suffer from feelings of guilt,

despite his unenviable track record. This was especially so early on in his lechery, as witnessed when he had a liaison with Mrs Jean Heron, the 17-year-old wife of a close friend. He was so mortified by his behaviour he took up with two middle-aged actresses to dilute the feeling.

When he caught the dose from Louisa he retired to his room to live on bread and water and take physic for five weeks. A friend of his father's visited him and left with a stinging pun, "Who in the performance of the manly parts would not wish to get claps?"

The profligate life continued, and Boswell in turn continued to feel self-disgust, not so much at the immorality of his behaviour but at its brutishness. But the loathing was not strong enough to make him stop: the masochism would appear to have been as pleasurable as the conquest.

He tells us that he longed for a "genteel girl" as a mistress. But before such a paragon could materialise, in June 1763 he met Dr Johnson. He at once admired the larger than life *bon vivant* and it was to him that he confided his sexual problems. As can be imagined, in reply the proffered advice was stern and morally uplifting. It may not have stopped Boswell in his activities, but it did stiffen his moral fibre.

Later that year Boswell went on the Grand Tour of Europe where he had "many victories", including Rousseau's mistress who told him, "you are a hardy and vigorous lover, but you have no art".

In Dublin he caught a dose of gonorrhoea which was so severe that he had to return to London for the cure. This consisted of camphor liniment and mercury plaster on the affected parts, a daily draught of a couple of pints of Kennedy's Lisbon Diet Drink (at half a guinea a bottle and a recommended dose of two bottles daily, it was clearly for the affluent lecher; he must vainly have willed it to work) and some minor surgery, presumably dilatation.

In a moment of inspiration and uncharacteristic perspicacity in 1769, he married Margaret Montgomerie. Though she had neither wealth nor title, she loved him, she understood him and above all she was tolerant of his quirks and foibles. Indeed, she possessed treasures which made her worth above that of rubies. And he knew it. In his own odd way he loved her in return and paid her the compliment of (almost) always being frank with her regarding his backslidings. They lived in Edinburgh where Boswell had a law practice, but he often used to venture up to London to meet Johnson and fall to the old temptations of drinking and womanising. He had had 10 attacks of gonorrhoea before his marriage, but only two between 1769 and 1779, something of a record for him.

James Boswell kept a very detailed diary, not only recording the famous bon mots of his mentor but also details of his own sexual outings. He does not seem to have visited James Graham's well-known Temple of Love and its celestial bed, or been known to have met Emma Hart (later Emma Hamilton and paramour of Admiral Lord Nelson), who is said to have worked in the establishment. He probably had no need of such bogus stimulants of the flesh, and while in town sought out more intellectual stimulants such as Samuel Johnson, Sir Joshua Reynolds, Edmund Burke, Oliver Goldsmith, Alexander Pope and other literary giants of the day, probably at the Cheshire Cheese in Fleet Street.

Margaret died in January 1789 while Boswell was away yet again in London trying to shape a legal or political or literary career. Her death depressed him and he slowly became a habitual drunken lecher, a familiar figure lurching through the less reputable streets of the capital. More attacks of his old trouble occurred, and he once had total stricture of the urinary outlet necessitating surgical intervention.

The vulgar debauchery continued until May 1795 when he eventually succumbed to, it was said, a bladder tumour, although complications of his chronic disease would surely have played a part. Furthermore, the mental degeneration of the last few years may have been due in part at least to repeated ingestion of the mercury pills he took for his condition.

As a man he had two endearing qualities. The first was his capacity to make friends; he seems to have had boundless good humour, a characteristic which would stand him in good stead in dealing with the demimonde of the seamy side of his life. The other was his intrinsic honesty, so that we know from his jottings more of the quirks and shortcomings of this man than perhaps anyone else in history. He spared his diary nothing.

His libido had been liberated by his rejection of a strict Calvinistic upbringing. That childhood, however, had created in him a sexual behaviour pattern which oscillated between self-reproach and swaggering machismo. Boswell was not a reticent man, but whatever his mood the recordings of it bear a ring of truth. It is perhaps this characteristic which eventually brought forth his great and polished literary product, his *Life of Johnson*.

He was also, of course, promiscuous even by the standards of the freewheeling times in which he lived. Gonorrhoea was the greatest risk and he contracted it on numerous occasions. In itself this had serious physical consequences, but more than that, the treatment was almost as dangerous as the complaint.

The disease was so widespread in the eighteenth century that the commonest advertisements in the contemporary periodicals were concerned with venereal disease, cosmetics and books in that order. Hunter's Restorative Balsamic Pills were claimed to restore and reinvigorate the constitutions of persons weakened by a course of dissipated pleasures. The Bath Restorative was for "those who have been almost worn out by women or wine".

Cures for venereal diseases bore such exotic titles that they almost invited the simpler souls to catch the disease in order to try them — Lisbon Diet Drink, Dr Solander's Vegetable Juice, The Specifi, and last but not least, Leake's Pills The Best Known Cure for Gonorrhoea, which were depicted in a Rowlandson cartoon of 1786.

Boswell also used condoms as a protection against disease rather than as a contraception, and they were purchased at Mrs Phillips's shop, The Green Canister, in Half Moon Street, Leicester Fields.

Eighteenth-century England was a rip-roaring world of abject poverty, drunkenness, political graft and loose sexual morals with accompanying social disease. But out of this miasma grew many towering intellects, a cultivated and leisured society for those with money and status, and for those without, an industrial revolution which changed the world.

Boswell was a part of all this, and whatever one may think of his licentious ways, his passionate and introspective honesty has given us a unique insight into a medical condition which is even now more known for its furtiveness and clandestinity than its ventilation.

Besides being a libertine, of course, James Boswell produced a biography of such scholarly proportions that it is still held as the yardstick by which similar works are measured. If he had thought of it, he could, in the end, have anticipated Hilaire Belloc and written :

When I am dead, I hope it may be said:
"His sins were scarlet, but his books were read."

(JL)

Chapter 5

Opium

SHAMEFUL START TO BRITISH RULE IN HONG KONG

Not long ago the British were lamenting the return of Hong Kong to Chinese rule. But they prefer to forget how their forefathers first acquired the colony. It was a shameful, sordid affair, based on hypocrisy, greed and the exploitation of opium addiction.

The opium wars of the nineteenth century were, according to one writer, "precipitated by the Chinese government's effort to suppress a pernicious contraband trade in opium, concluded by the superior firepower of British warships, and followed by humiliating treaties that gave Westerners special privileges in China".

How did all this come about?

Way back in 1599, Queen Elizabeth I granted 218 London merchants monopoly rights over all trade in the East Indies (south-eastern Asia). So started the East India Company, which conquered and governed much of India as a company colony. During the eighteenth century, Londoners were drinking so much tea (mostly from

China) that they spent about 5% of their earnings on it. The government was equally dependent on excise duties from tea. By the early 1800s, the East India Company was financing its tea trade by illegal exports of opium to China.

For over a thousand years, the Chinese had used opium, but mainly for medicinal pruposes. Then, during the eighteenth century, recreational smoking spread. The Chinese government passed edicts that proved to be as ineffective as Prohibition in the United States.

Pressure for free trade ended the East India Company's monopoly in 1834. Soon a vast network was moving opium, even into northern China. Missionaries described the effects:

Those who begin its use at twenty may expect to die at thirty ... a frightful ... atrophy reduces the victim to a ghastly spectacle who has ceased to live before he has ceased to exist.

Since opium provided one-sixth of Britain's national revenue, the British government supported the trade. No wonder the Chinese loathed the British. But the British said they were just meeting the demand. Never mind that they encouraged addiction by distributing free opium, and bribed police to turn a blind eye.

The Emperor of China proclaimed the death penalty for Chinese opium growers, distributors and users, and even for foreign importers. He sent to Canton Commissioner Lin, who had reduced opium problems in central China. Lin offered addicts who gave up opium medical care to ease their withdrawal. He wrote futile letters to the British ruler, asking for the shipments to stop. But in Canton, Lin smashed the domestic opium network; by early 1839, the trade seemed dead.

Surrounding the foreign factory area with troops, Lin demanded the surrender of the opium stocks. To protect the 350 European and American traders there, Captain Elliot, the British superintendent of trade (who himself despised the trade), promised to indemnify British dealers who complied. Elliot's expensive capitulation did not amuse Queen Victoria. While destroying the

The British developed a very lucrative drug business that would have put the Colombian drug lords to shame ... all at the unfortunate expense of the good Chinese folk.

confiscated 1000 tons of opium (which took 23 days), Lin beheaded a coolie who tried to sneak a little for himself. But the British dealers refused Lin's demand to sign bonds promising never to trade in opium again.

In July 1839, after drunken British and American sailors assaulted a Chinese man who died, Elliot refused to surrender them to Chinese justice. Instead, he himself tried them, found them guilty, but only jailed and fined them. This leniency infuriated Lin.

So started the still undeclared Opium War of 1839–1842. In September, Elliot fired on Chinese forces in Kowloon. In November, the British sank four Chinese war junks. The first phase of the war brought only a short-lived peace. In January 1841, Elliot and Qishan, the new commissioner replacing Lin, reached an agreement, which both their masters promptly repudiated.

Foreign secretary Palmerston blamed Elliot for accepting too small an indemnity, which included the rocky, arid island of Hong Kong! Palmerston had no inkling of the enormous potential of

Hong Kong as a free port, which it became in 1841. The Emperor was furious that Qishan had yielded any Chinese territory at all; Qishan returned to Peking (Beijing) in chains.

For the second phase of the war, from 1842, the British had a new leader, the daring veteran Sir Henry Pottinger. Pre-industrial China had no hope against British muskets, heavy artillery and paddlewheel gunboats. By August 1842, the British ships had reached the outskirts of the southern capital Nanking (Nanjing) and prepared to shell the city.

About 10 000 British had overcome an empire of 370 million. The Treaty of Nanking opened China to the West and marked the beginning of a century of foreign exploitation. This was the start of the humiliating "unequal treaties" imposed on China first by Britain, then the USA, and then France.

Surprisingly, the treaties said nothing at all about opium. Not surprisingly, the British continued to import opium illegally. In Britain, the humanitarian Lord Ashley spoke out against "one of the most lawless, unnecessary and unfair struggles in the records of history".

Once opium in China became legal in 1856, smuggling quickly ceased. Though opium remained a serious social problem, the edicts forbidding its use were not revived until 1906. In the 1920s, regional warlords in China started growing opium to finance their armies. Even now, smuggling from the Golden Triangle into China remains a major problem.

(GB)

Taking opium was like playing Russian roulette

More than three centuries ago a famous physician declared, "Among the remedies which it has pleased Almighty God to give to man to relieve his sufferings, none is so universal and so efficacious as opium." In those days a popular remedy was laudanum (opium mixed with alcohol). Its users were actually getting two fixes.

Even in the late eighteenth century, pharmacists still got their drugs from plants. But, like other plant extracts, opium had one major downside: the varying strength made its effect variable. Sometimes it had little effect, other times it killed people. Modern chemistry was only in its infancy. When its founder Antoine Lavoisier went to the guillotine in 1794, Friedrich Serturner was still a boy in Westphalia, Germany. His family was poor, and things got worse after his father died.

At 16, he became apprenticed to the royal pharmacist. Shuttling between the pharmacy and his improvised basement laboratory, Serturner started experimenting. Frustrated by the unpredictability of opium, he took on a huge task: to find its active ingredient. After 57 experiments over four years, he managed to isolate the active component (bitter, odourless crystals of morphine). Now Serturner found that adding it to the food of mice and stray dogs put them to sleep; larger doses killed them.

But he went beyond animal tests and tried it on himself and three 17-year-old friends. They were lucky to survive.

First, each took 30 mg (a large dose) of morphine, dissolved in alcohol and water. They felt flushed and feverish. After half an hour, they boldly repeated the dose; they became nauseated and dizzy as well. After 15 minutes, they took their third dose. The young men got sharp pain in their stomachs and felt faint. Serturner lay down and dozed off. Luckily, when he woke he had his wits about him; he gave everyone enough vinegar to make them vomit violently. One boy was still seriously ill, so Serturner also gave him carbonate of magnesia. For some days, all four suffered stomach pains.

In 1817, he named his active principle morphine, after Morpheus, the Greek god of dreams. In the long run, this discovery pointed the way to precise, predictable dosages of pure drugs. But when he published his work, the journal editor fiercely attacked him. Serturner responded to criticisms by doing further meticulous experiments.

Controversy dogged Serturner for the rest of his life. Fourteen years before he had isolated morphine, a French pharmacist had done similar work, though his extracts were very crude. There followed the usual disputes about priority and even charges of plagiarism. Eventually, the Institut de France awarded Serturner a prize of 2000 francs for his discovery.

It was fitting that morphine eased the agonies of the gout that gradually immobilised Serturner himself.

His was a giant intellect ranging far beyond his own specialty. He spread himself over chemistry, physics and even medicine. Before the pioneer work of Pasteur and Koch on germs that caused infectious diseases, Serturner described the cause of cholera as a poisonous, animated being that can reproduce itself. When a cholera epidemic hit Europe in 1831, he printed pamphlets at his own expense, urging people to avoid cholera by boiling drinking water. Millions of those who failed to heed Serturner's advice died of cholera.

Though Serturner isolated morphine in 1803, it was not until the twentieth century that chemistry advanced far enough to determine its chemical structure. In 1952, chemists finally synthesised morphine.

In 1841, Serturner collapsed and died while drinking tea. Surely if the Nobel Prize had existed then, he would have won it. One biographer wrote: "Serturner's discovery stands well alongside the greatest discoveries which have benefited the human race."

The story of new, powerful painkillers rolls on. Serturner started with opium from which he derived morphine; later, he warned people that morphine was addictive. Later still heroin (derived from morphine) came into use as the non-addictive answer to morphine addiction. In 1900, Bayer Pharmaceuticals was advertising heroin as the "sedative for coughs".

Now about 500 Australians are dying each year of heroin overdoses. But we cannot put the genie back in the bottle.

(GB)

Chapter 6

FAMOUS AND INFAMOUS PEOPLE

ATATURK AND THE ANZACS

We are Aussies on a tour of Gallipoli. Geoff has come on a pilgrimage to the grave of his grandfather. He and our Turkish guide Ali share a bond: Ali also lost his grandfather here. My eyes grow misty as I read the inscription on the monument:

There is no difference between the Johnnies and the Mehmets where they lie side by side ...
Your sons are now lying in our bosom and are at peace ...
They have become our sons also.

The last words are "Ataturk 1934". Who was this Ataturk? Mustafa Kemal, not yet known as Ataturk, was born to an undistinguished family in 1881 in Macedon, then part of Turkey. Though arrogant, prickly, stubborn and rebellious, he excelled at military school. He was able, ambitious and fearless. Turkey was "the sick man of

Europe". Russia had taken the Crimea, France had annexed Algeria, the British had taken Egypt and Greece had taken Crete.

The corrupt, despotic sultan ruled from Constantinople; his spies were everywhere. Arrested as a leader of a revolutionary group, Kemal languished in solitary confinement and was nearly executed. He took up his lifelong mission: to deliver Turkey from what he saw as its tyrannical rulers, obsolete customs, illiteracy and Muslim conservatism.

During the First World War, Turkey allied itself with Germany. In April 1915, British, French, Australian and New Zealand troops landed at Gallipoli, where Kemal, now a general, had charge of a half-formed division. Ignoring orders, he threw every available Turk into a desperate defence. Despite an attack of malaria, he led his troops from the front throughout three sleepless days and nights. Kemal escaped death when the watch he carried in a breast pocket blocked a piece of shrapnel.

Altogether, on both sides, there were about half a million casualties at Gallipoli; that is, incredibly, half of those who fought there!

Some accounts state that deception by the Anzacs kept the Turks unaware of their withdrawal from Gallipoli and so prevented further casualties. But our guide Ali insists that the trenches of the two sides were far too close for such deception. The Turks, he says, had come to respect the Anzacs and would not shoot at withdrawing men. He tells us a tale of fraternisation at Gallipoli.

One day late in 1915 the Turks dropped a bag of tobacco into the Australian trenches with a note: "We have tobacco, you have paper." Next day, the Aussies threw back half the tobacco, rolled up in odd bits of paper: old newspapers, even letters from home.

Kemal's defence at Gallipoli made him famous as "The Saviour of the Dardanelles". But Turkey was in terrible shape. Famine and disease were widespread. The politicians were obeying the allied orders to disarm. They in turn ordered Kemal to demobilise the remaining Turkish forces on the Black Sea coast. Instead, in May 1919, Kemal set about inspiring and organising them.

In August 1920, the helpless sultan signed the Treaty of Sevres. This capitulation would have reduced Turkey to a puppet state at the mercy of its neighbours, but the nation rallied behind Kemal and rejected it. In 1922, his nationalist forces finally drove the Greek invaders from Turkey. It is said that one disturbed Greek general stayed in bed, since he believed his legs were made of glass and would break if he got up!

The National Assembly abolished the office of sultan. Russia, France and Italy signed the Treaty of Lausanne, which set Turkey's borders about where they are now. On 20 October 1923 Turkey became a republic, Ankara its new capital and Kemal its president.

Turkey, he declared, must adopt Western dress, laws, education and constitution. Kemal made Turkey a secular state and did away with Muslim schools, the Islamic legal system, and the wearing of veils by women and the fez by men. He cleaned up the corrupt civil service and granted women the right to vote and to hold public office. When he made family names compulsory, the assembly granted him the surname Ataturk (Father of the Turks).

Kemal controlled the assembly and could appoint and dismiss the prime minister and cabinet. He censored newspapers, and tortured and hanged hundreds of Turks. Police tearing fezzes from Muslim heads led to riots and the stoning of officials. Religious leaders denounced him. Kemal savagely put down a Kurdish revolt in 1925.

Vivian Green attributes his ruthlessness to psychosis (mental disturbance) caused by a lack of the vitamin thiamine in his poor diet brought about by alcoholism. Certainly, Kemal did not hide his heavy off-duty drinking. A French journalist wrote that Turkey was governed by one drunkard, one deaf man (Kemal's prime minister) and 300 deaf-mutes (the deputies). Kemal replied: "This man is mistaken. Turkey is governed by one drunkard."

By 1937, Kemal was very ill. When he complained of an itch, officials started a futile search of the presidential palace for the insects that must surely be biting him. His own doctors failed to diagnose his illness, though he developed the swollen belly and jaundiced skin of terminal liver failure.

On 10 November 1938, Kemal Ataturk died in the Dolmabahce Palace in Istanbul, where the clocks still remain stopped at 9.05 a.m., the time of his death.

Even now, more than 60 years later, Turks honour him with his likeness on banknotes, stamps, statues and portraits. Ambitious politicians still claim to be the inheritors of his mantle. History will record Kemal as a ruthless dictator, but also as the greatest figure in modern Turkish history. His monument is modern Turkey itself.

(GB)

KING EDWARD VII:
A ROYAL BON VIVANT

On 6 May 1910 George, eldest son of the reigning monarch of Great Britain and her dominions across the sea, including Australia, bent down close to the puce-coloured face of his dying father and whispered not that the Empire was safe, as you may think, but that the king's horse "Witch of the Air" had won that afternoon at the Kempton Park racecourse. The breathless old man managed a thin smile and gasped in reply, "I'm very glad." Hardly memorable as last words go, they were to be his final pronouncement before succumbing shortly afterwards to a final heart attack following on from his severe bronchitis.

The son was later to say, "I have lost my best friend and best of fathers. I am heartbroken and overwhelmed with grief." I am sure he was, but they had not always seen eye to eye, and indeed his father, Edward VII, had led a strange, dissolute life seemingly at variance with both that of his mother Queen Victoria (and who more Victorian than she?), and of his rather inflexible son, who became George V. This could well have been a result of the strained relationship that existed between himself and his mother, and the general lifestyle of his social set.

So let us look first at his lifestyle, the medical high point of which was the stopping of his coronation for good clinical reasons, and then turn to his well-known sexual athleticism, when the appellation "Father of the Nation" took on a whole new meaning.

Albert Edward was born in London in 1841 and was the second child and first son of Queen Victoria and her consort, Prince Albert. The elder child, Victoria or Vicki as she was always known, had been a difficult labour, and the Queen looked forward to the second pregnancy with ill-concealed apprehension. She was not to be disappointed, for during it she wrote that she felt "very wretched, low and depressed". The actual labour was accompanied by a suffering which "was really very great". As well as a distaste for labour itself, Victoria had a well-known dislike for small children. So all in all there must have been a general air of gloom and impatience at the palace.

As a boy Albert Edward was known as "Boy" or "Bertie". In later life he came to be known under several sobriquets from "Kingy", "Tum Tum" to "Edward the Caresser" and other even more scurrilous nicknames. But we're getting ahead of ourselves here.

His early education was undertaken by an old-timer who had himself been educated during the French Revolution and was quite out of touch with the general mores of the day. He must have been interesting to talk to if you could steer him away from Euclid or Virgil, a ploy quickly spotted by such tearaways as Bertie and his brother. Victoria herself kept Edward in the dark as far as state matters were concerned, despite the fact that he was heir to the throne, and that he wished to be involved.

So here we have a set of circumstances of a loathed pregnancy, a strict and inept education, and a spurned enthusiasm on the one hand, and youthful boisterousness, wealth and persuasive friends on the other. Well, what would you have done? He did the same. Psychologists could have a field day explaining the events of his adult years, but whatever the reason he led the fascinating life of a bon vivant and rake.

Let's start with his eating habits. Because there is no stronger label, they could only be described as gluttonous. Admittedly, it was an age of the hearty appetite, but like a true king, Edward led from the front. He devoured five solid meals a day (tea and supper were both full meals). He started with a glass of milk in bed. Up to toast and coffee, and just before the morning shoot platefuls of bacon, egg, haddock, chicken, toast and coffee. He ate rapidly and with relish. The fresh air sharpened his appetite and a bowl of soup was put away mid-morning. A multi-course lunch at 2.30 p.m. would not put him off having tea two hours later. The band played at this indulgence as he tucked into poached eggs, petits fours and preserved ginger, hot cakes, scones and his favourite, Scottish shortcakes.

Dinner followed at 8.30 p.m. and was, of course, the main meal of the day for the by-now ravenous monarch. It comprised 12 courses and he usually never missed a beat. Oysters were followed by maybe plover's eggs, poached sole, turkey in aspic, quail pie, grouse, snipe, woodcock and other birds he had earlier blasted out of the skies in their hundreds. The thicker the dressing, the richer the stuffing, the creamier the sauce, the greater the enjoyment. He never grew tired of partridge stuffed with truffles or of boned pheasant laced with paté de foie gras. The only thing he did not like was macaroni, perhaps because it had not been shot, reeled in or hunted to exhaustion.

The edge of his appetite was not dulled by his smoking habits. By the time he had reached breakfast the King had had two Egyptian cigarettes and one cigar. By the end of the day he had got through 20 cigarettes and about 12 Corona y Corona cigars, his favourite.

If smoking ever blunted his appetite, which it did not, it could be revived again by a snifter or two. In his younger days he used to decant a bottle of champagne into a jug and help himself during the meal. When he became king he rarely had more than two glasses. He hardly drank wine and finished the meal with a cognac. Edward may have had several vices, but heavy drinking was not one of them.

Like many who guzzle their food, he ate quickly and was resentful if conversation interrupted the relentless shovelling down of food. The story is told of his grandson, the future Edward VIII, beginning to speak during a meal to be instantly silenced by the gruff king. As the plates were being taken away the boy was asked what it was he had wished to say. The reply came back, "It's too late now, Grandpa. It was a caterpillar on your lettuce, but you've eaten it."

He was impatient with people who ate slowly, so when he finished everyone put their cutlery down. As a result no-one left the best bits till last, and many a lifelong habit had to be broken.

Prince Edward's feasting was common knowledge and a source of secret wonder. Eventually Parisian nightclub singers began to mock the regal gourmandising, so it was decided that, as a sop to public disdain, for a short time each year a diet be undertaken at such desirable and chic places as Marienbad or Biarritz. If you are going to attempt the impossible, it might as well be done in style.

Bertie drank the mineral waters with as much grace as he could muster and did away with the half-chicken which customarily stood on his bedside table in case he awoke peckish in the night. But the local hoteliers in these parts did not wish to appear niggardly to a visiting monarch, and, by force of habit when in the presence of royalty, put little culinary temptations in his way in the form of goose stuffed with aubergines and other calorie-laden snippets. Their meals seemed to be the epitome of luxurious habits instead of a disincentive to overeat. The most the king ever lost was three and a half kilos in two weeks. I suppose even that was a minor triumph in view of the irresolute way the problem was tackled. Paradoxically, he was weighed every day by the local cake-shop proprietress. I wonder where her interests lay.

There can be no doubt, of course, that his overindulgence affected his health and feeling of wellbeing. After all, he was only human. So let's consider his health.

In 1871 Bertie had an illness which strangely enough helped bolster the flagging interest in the royal family in England. Since the death from typhoid of "Dear Albert", the Prince Consort, in 1861, Queen Victoria had retired from public life. After ten years the populace not unreasonably thought that enough was enough, and there were rumblings about the country of getting rid of the unseen monarch. By what in retrospect was a most fortunate if somewhat serious illness, Bertie's recovery lead to a warm burst of royal approbation and no more anti-monarchist murmurings were heard.

It seems that in late October 1871 the prince was a house guest at the country residence of the Earl and Countess Londesborough. A few days after his return to Sandringham, the royal residence in Norfolk, the prince fell ill. Ten days later, on 23 November, it was announced that he had typhoid fever. Evidently, the Londesboroughs' drains were not on a par with their hospitality. A week later, to everyone's horror, a fellow guest, the Earl of Chesterfield, died of the same disease. Worse, the prince's groom followed Chesterfield to the cemetery. It was known that Edward's father had died of the self-same malady, so by now the court had the breeze up.

On 29 November the royal patient began to rave and shout indiscretions in his ramblings. As names were being named, the Princess of Wales was kept out to save unnecessary embarrassment all round. She divided most of her time between praying, hand-wringing and having attacks of the vapours. That day Queen Victoria herself arrived to add to the social complications.

By 7 December it seemed all hope of recovery had gone. A day of national prayer was declared and five bulletins were issued over 24 hours. But the delirium eased, and, as the Queen was to write in her diary, "he was brought back from the very verge of the grave". On 15 December Edward smiled at the assembled Jonahs and asked for a glass of Bass's beer. He took some weeks to recover completely and finally receive the warm acclaim of the people.

In 1895 Prince Albert Edward developed rheumatism in his right shoulder brought on by the vigour with which countless thousands of loyal subjects shook his hand. On account of this he devised "The Prince of Wales" handshake, whereby he kept the elbow tucked into the chest wall. Being the leader of society that he was, he changed the fashion for formal recognition as people adopted this rather stiff greeting. Many years later his grandson, Edward VIII, suffered hand pain from the same thing and took to using the left hand for such courtesy.

But the illness for which Edward VII is best remembered, of course, is an attack of appendicitis a couple of days before his coronation.

By the time his number came up to be king he was balding and middle-aged, and doubtless wondered if the moment would ever arrive, or if he would die first of overindulgence, boredom or sheer frustration. He planned it with much joy and nitpicking detail for it was the first such event for 64 years.

Suddenly here it was; 26 June 1902 was the appointed day.

May was particularly wet that year, even by British standards, and on 11 June the uncrowned heir was said to be suffering from a chill. He missed two of his favourite outings, a military parade at Aldershot and the horse racing at Royal Ascot, so something must have been wrong. He elected to rest at Windsor, but neither the court nor the general public thought too much of it.

But the inner circle had noted a few changes. After the so-called chill, there had been loss of appetite, a very significant symptom in this particular patient. His drinking, on the other hand, increased and he became irritable and sleepy. On 23 June he returned to Buckingham Palace, insisting that the plans were to go ahead. That night he developed severe pain in the lower abdomen and despite his at that time 48-inch waist, his doctor Sir Francis Laking was brave enough to diagnose perityphlitis, as acute appendicitis was then called.

At that time operative treatment for this condition was in its infancy; indeed, the diagnosis was a new trend and the pathology poorly understood. It had usually been treated by diet and carminatives. The mortality rate was uncomfortably high. Laking was on the horns of a dilemma. He knew the risk of operating on the one hand and of conservative treatment on the other, and also knew there was not a lot in it: both could be pretty disastrous.

Further, he knew, poor devil, the eminence of his patient, his temper, and the ominous ceremony ahead. He came down on the side of a milk diet and bed rest. And who can blame him!

The patient got worse and peritonitis set in. Even so, Edward was unrelenting in the way he fretted about the details of the forthcoming celebrations. He even called in a gypsy to forecast the outcome and was alarmed at the given prognostication that he would not be crowned, and, worse, her own imminent death would soon be followed by his. She did, in fact, die about a week later but must have told the right forecast to the wrong man. Sir Francis Laking called in Sir Thomas Barlow and both now agreed that operative intervention was essential. Without asking the authorities, they alerted yet another medical knight, Sir Frederick Treves, to lay out his best suit and sharpen his scalpel, and had a room in Buckingham Palace set up as a theatre.

The distinguished patient was approached. Operation was essential, or death would follow, as night follows day. The heir was intransigent. "Laking," he said, "I will stand no more of this. Leave the room at once." Barlow left, but the valiant Laking stayed to tough it out. At last a compromise was reached: Treves would be seen; no promises, mind. The smooth-talking surgeon succeeded where the others had failed, and at noon the next day Edward walked into the makeshift operating theatre wearing his old dressing gown and carpet slippers. I wonder if he gave a thought to his predecessor Charles I walking to the scaffold about 250 years previously: Edward's chances were only marginally better.

Queen Alexandra helped hold him down as he bucked and plunged under the chloroform. Treves had to ask her to leave so that he could roll up his sleeves and don his apron for the actual cutting. Sterile masks and gowns were a thing of the future, just as old-world charm was a thing of the present.

The operation took forty minutes, and as the anaesthetic wore off the King opened his eyes and said, "Where's George?" George, Prince of Wales, saw him the next morning when he found him sitting up in bed smoking a cigar. When the Queen visited he pretended to be asleep, as he found shouting to counter her deafness hurt his wound.

Laking and Treves were created baronets, and I would say thoroughly deserved the accolade. Indeed, rabbiting away through the 48-inch waistline of the foremost in the land is probably worth an earldom plus a week of rough shooting at Balmoral at the very least.

The king was declared out of danger on 5 July and the coronation eventually took place on 9 August 1902.

During his reign the king went on his merry gourmandising way, but by now he was in his 60s and the good life was having an impact. He began to get rheumatism in the knees, a situation not helped by his excess weight. In 1909 on a visit to Germany, and while trussed up in the high-necked Prussian uniform which protocol demanded, he had a fainting fit. The dear lady whose hand he was pressing at the time thought he had died and had herself to be revived. On that same trip the locals noted that the king was so obese that he gasped and choked as he went upstairs, and the effects of chronic bronchitis began to manifest themselves with a morning cough and some wheezy breathlessness.

Sometimes he spluttered incessantly all day. He had frequent sore throats and bouts of fatigue. Yet he could not be persuaded to stop his obnoxious cigars, the puffing of which seemed to be as prominent a part of his appearance as his straining waistcoat buttons, ample thighs and hooded eyes. While he was prickly with his doctors and dismissive of their advice, he actually became depressed with the stresses of office and fearful of old age.

In February 1910 the king suffered an attack of sudden chest pain and breathlessness. Could it have been an unrecognised coronary thrombosis? Whatever it was, he recovered but in April again came down with bronchitis. On 2 May he spent several hours in the open sitting astride his horse in inclement weather. And on 6 May, as we have seen, he died.

But he was a merry monarch and I do not think he would have liked us to have taken our leave of him on his death bed. It is more fitting that we should salute his passing with an irreverent review of that part of his life for which in later years he has become most well known: not the Entente Cordiale, not his freelance diplomacy, not even his epicurism, but his legendary sexual dynamism.

Throughout his life Edward was a rake of Catholic tastes and as prepared to sleep with a fashionable harlot as a duchess. More than that, he was a rake who loved the rakish world in which he moved. He was untrammelled by finer feelings or doubts about his life. His girth was no bar to his exploits, possibly because it was regarded as an honour to be invited into his boudoir, possibly because it was regarded as bad taste to mention any shortcomings that may have been experienced on account of it. He himself seemed to regard it as his birthright and early on commented, "High station has, after all, some merits, some advantages."

He doggedly pursued his libidinous habits which became a rip-roaring way of life and an outlet for his considerable energy and a prop to shore up his otherwise rather meaningless existence.

It is said that he lost his virginity in 1861 at the age of 20 while in Ireland at a military camp. The story goes that his madcap fellow officers persuaded a notoriously promiscuous actress called Nellie Clifden to climb through his window or under his tent flap (accounts vary) and insinuate herself into his bed. A jape like that was soon the talk of the mess, and it eventually came to the ears of a sickened Prince Consort. There was a parental showdown of monumental proportions and not long afterward Albert died. Queen Victoria blamed her eldest son for the death (in fact, as we have seen, Albert died of typhoid) and wrote to Princess Victoria, "I never can or shall look at him without a shudder. He does not know that I know all — Beloved Papa told him I could not be told all the disgusting details ..."

Well, whatever the disgusting details were, bold and brazen Miss Clifden had opened up a whole new world for Bertie, and one he basked in for the rest of his life.

He found that in his role of heir to the throne he had extraordinary licence and was presented with far more sexual opportunities than was good for a young lad or he was able to cope with. Rumour gathered, as rumour has a habit of doing, that he was one of the generation's lecherous greats, a byword in stamina and strength. It was said that for a husband it was more duty than dishonour to allow a wife the chance of royal, nay regal, pleasure. Well, maybe.

He was not into young, unmarried girls — too many potential complications. But he did like variety, and, rather dangerously, excitement. He appeared to consume his women with the same gusto and lip-smacking enjoyment that he did his food. His basic *plat du jour* diet was from a menu of carefully chosen actresses. The à la carte were the married women of his fast set. These schemers were usually visited at their homes in the afternoon while their conniving husbands were at their clubs. By tacit consent nobody questioned the presence of the prince's cab parked in a Mayfair or Belgravia street. The main course, or *specialité de la maison*, was the current favourite or society beauty whose favours were fawned over in the joyous licence of a weekend house party or autumn shoot.

Bertie's invaluable and extremely discreet secretary, Francis Knollys, commonly arranged the various trysts and there was a general cover-up by society, particularly within his own clique, so no breath of scandal emerged. Indeed, at his coronation at Westminster Abbey in 1902 a special stall for his entourage was set aside. To those in the know it was euphemistically called "the King's loose box".

Among his well-known conquests were "the divine" Sarah Bernhardt (I wonder if her wooden leg added special piquancy to a jaded sexual palate); "the Jersey Lily" or Lillie Langtry; Catherine Walters, also known as "Skittles", one of the best known high-class prostitutes of the Victorian era; the Countess of Warwick, also known as "Darling Daisy"; and Mrs George Keppel, the last of a mostly graceful and elegant line. In a pleasing act of contrition Queen Alexandra allowed Mrs Keppel a few private minutes with Edward VII when he was on his death bed. There were many others, and today visitors to England are constantly pointed out country houses said to be the scene of Edwardian sybaritism.

Queen Alexandra knew of all the royal goings on, and, whilst not approving, turned a deaf ear. This is not so fanciful as it sounds, for she was in fact quite hard of hearing, a disability which worsened with age. As a child she had had tubercular glands of the neck incised. This procedure had left a scar and in later years she hid it by the wearing of a high jewelled "dog collar", thereby setting a fashion which was to last for years. She was also taller than the king. Whatever else she may have been, such as a chronically poor time-keeper, Alex was loved by Bertie and she in turn was loyal to him. There was never any question of separation.

Edward was entangled in one scandal from which he derived some public sympathy. This was the Mordaunt case in which Sir Charles Mordaunt brought a divorce case against his wife, who by then was in a mental institution. It seemed that the prince had corresponded with this lady in happier times, and he was called as a witness. To his mortification his letters were read out in court. It was all very embarrassing, but no stigma appears to have been attached to him.

His widowed and imperial mother was not so easily placated and wrote, "...the Prince of Wales's intimate acquaintance with a young married woman being publicly proclaimed will show an amount of imprudence which cannot but damage him in the eyes of the middle and lower classes, which is to be lamented in these days when the higher classes, in their frivolous, selfish and self-seeking lives, do more to increase the spirit of democracy than anything else". Tortuous though the sentiments are, I think I know what she means. But the fact remains that it did not do Edward much harm in the country's eyes. Perhaps people just did not care.

But when all is said and done, Albert Edward had a single overwhelming talent — an ability to put people at ease. He was likeable and personable and, of course, had an unwavering eye for a pretty woman. From a medical point of view he did everything wrong. He smoked too much, he ate too much, he paid scant attention to bodily ailments, and apart from riding he took little exercise. He was popular and led English society to the manner born so that any affectation such as using a walking stick, wearing a homburg hat or having his appendix removed gained instant respectability and set a fashion.

Looking at him now, a century on, the capacity to enjoy himself in a completely unabashed way is still as refreshing as it was contemporaneously unique; a characteristic worthy of a king.

(JL)

ELVIS PRESLEY: WHAT KILLED THE KING?

As the ambulance pulled up, a swarm of men surrounded the stretcher. Hospital staff worked frantically until at last a young nurse demanded, "Why are we still working on this corpse?"

"Because he's Elvis Presley."

That was August 1977. Elvis was only 42 years old.

Elvis and his stillborn twin brother were born in 1935 in a two-roomed shack in Mississippi. As a toddler, Elvis first sang along with the choir of the Church of God. Later, his manager, "Colonel" Tom Parker, boasted: "When I met him, he only had a million dollars worth of talent. Now he's got a million dollars."

In 1956, Elvis recorded *Heartbreak Hotel*, the first of his 45 records that sold over one million copies. He also made his first film, *Love Me Tender*. But it was live shows that made his name. Critics accused his hip-shaking "Elvis the Pelvis" style of "sexually setting young women on fire".

Under all the hype, wealth and glitter, the real Elvis was a vulnerable man shattered by his mother's early death. To fill the void inside him, he tried flash cars, karate, religion, marriage and fatherhood. He sometimes gave away a Cadillac to a perfect stranger. But none of this satisfied him; nor did all the women he kept bedding.

Several doctors pandered to his growing drug dependencies: uppers before a show, downers to sleep and testosterone for a

Elvis "The Pelvis" Presley!
... seen here prior to his 1956 success, researching
possible material for future works.

flagging sex drive. With massive doses of cough medicine (Hycodan), he and a playmate almost overdosed in bed. Elvis also liked to play with guns. While stoned, he was a double menace and once nearly shot a girlfriend. Towards the end, he became paranoid, not only about people, but also about germs. His minders ("the Memphis Mafia") covered up whatever they could. When Elvis missed out on playing James Dean in a film, he blamed them; by this time he weighed over 110 kilos. Finally, his wife Priscilla divorced him. In December 1976, he wrote: "I feel so alone sometimes ... Help me, Lord."

When Elvis died on 16 August 1977, hundreds of thousands mourned outside his estate. One rumour had it that he had died of bone cancer, another that it was suicide or even murder. In the cemetery, police picked up three men carrying crowbars, wire cutters, shotguns and hand grenades! Suspicious people said they were about to steal the body and hold it for ransom. But they said they were fans who just wanted to open the casket. Why? To show that it was empty and so prove that Elvis still lived. Police dropped the charges.

But what did kill the King?

The medical examiner blamed high blood pressure and denied any evidence of drug abuse. But he would not release the autopsy report. After two years of investigation, the American Broadcasting Corporation went public. It turned out that a second set of blood samples had gone to an independent laboratory. On television, pathologist Dr Cyril Wecht said this new evidence showed that Elvis had died of an accidental overdose: codeine, Valium and a host of other drugs. In the seven months before Elvis died, he had taken more than 5300 tablets, which is over 25 a day.

One question remains: did Elvis deserve to be a superstar? A San Francisco journalist wrote: "That someone with so little ability became the most popular singer in history says something significant about our cultural standards." Bing Crosby had a bit each way, saying: "He [Elvis] never contributed a thing to music", but also "The things that he did ... that he created, are really something important."

In a way, the King still lives on. Over 20 years later, his fans console themselves with the host of Elvis impersonators who keep popping up. What's more, people continue to write books about him. One sums up his whole life in just four words. It's called *The World's Wealthiest Losers*.

(GB)

Was John Curtin a war casualty?

John Curtin, Franklin Roosevelt and Winston Churchill led their countries in the Second World War. But by 1944, they were so sick that no employer of today would have let them even push a pen.

Did poor health affect these men's capacity to lead? Conversely, did the pressures of leadership affect their health or even shorten their lives? In the case of John Curtin (1885–1945), my answers are yes, yes and yes again.

Curtin was the eldest of four children of poor Irish-born Catholic parents in Melbourne. Curtin started work as a copyboy at the *Age*. As well as playing football and cricket, he liked to read 'serious books' at the public library.

He joined the Victorian Socialist Party and in 1911 was appointed secretary of the Timber Workers' Union. Friends admired his intellect and idealism, but by 1914, they also worried about his drinking. Soon he resigned from his union post to convalesce from alcoholism. By 1916, he was drying out. After Curtin married and moved to Perth, he remained dry for most of the next ten years. He stood as a candidate for the seat of Fremantle at the 1919 Federal election, but lost badly. Exhausted and depressed, he had to take six months complete rest.

In 1927–28, his work as the editor of a union newspaper often took him away from Perth and he was drinking again. Even so, he won the marginal Federal seat of Fremantle. Despite his intellect and broad experience, Curtin did not become a minister under James Scullin. "For two years he was frustrated, underemployed, morose, lonely and drinking — at his worst."

In 1935, the ailing Scullin resigned from the federal Labor leadership. Curtin's colleagues gained his pledge to remain dry (which he apparently kept for the rest of his life) and surprised the country by electing him leader of the Australian Labor Party. In October 1941, Curtin became PM and also took on most of the wartime burden of defence.

When the Japanese attack on Pearl Harbor brought the United States into the war, Curtin declared war on Japan without consulting Britain. He said quite bluntly: "[For its defence], Australia looks to America, free of any pangs as to our traditional links or kinship with the United Kingdom." For an Australian statesman, this represented a quantum leap; Winston Churchill was outraged. Is it fanciful to see John Curtin as a forerunner of our present republicans?

Soon Curtin had to withstand more heavy pressure from Churchill who kept pressing to redirect the 7th Division from the Middle East to Burma. Curtin stood firm and brought our troops back to defend Australia.

Later in 1942, he won another bitter battle. Somehow Curtin was able to reverse the Labor policy that had prevented Australia from sending conscripts to fight overseas. This was a dramatic turnaround, not only for his party, but also for John Curtin himself. Way back in 1916, he had bitterly opposed conscription for overseas service and even gone to gaol for his beliefs.

When Arthur Calwell and other party traditionalists viciously attacked him on conscription, Curtin offered to resign. Caucus unanimously backed him, but the victory cost him dearly. A biographer wrote: "The strain on Curtin progressively increased." As well as his physical complaints, he was exhausted and depressed. He endured neverending travel by train or car to Melbourne and Sydney, as well as long trips over the Nullarbor. Why not fly? Because three leading members of the Menzies ministry had died in the Canberra crash of 1940.

His ailments included psoriasis (a skin complaint), pneumonia (in 1941) and, as noted above, episodes of depression. In April–June 1944, while in Washington to meet President Roosevelt, Curtin spent several days in bed with high blood pressure and neuritis. After his return, he became more irritable, touchy and lonely. Labor icon Dame Enid Lyons implored him to retire. "Easily tired, beset by endless problems his failing strength would not allow him to handle, he became irritable."

In November 1944, a heart attack kept him in hospital for two months. Historian David Day called this illness "the culmination of years of stress, heavy smoking, alcoholic binges and a simple but poor diet."

Returning to duty in late January 1945, Curtin could no longer cope; the government was hamstrung. Later one official complained: "It was impossible. Almost — almost — one could be glad he's dead so we can do things."

In April, Curtin's lungs became congested, but after several more weeks in hospital, he insisted on returning to the Lodge. Curtin himself said he was no longer fit for office, but a friend advised him: "Carry on quietly, and when you recover, then come to a decision."

He did not recover, but died in office on 5 July (only a few weeks before the war ended) "a war casualty if ever there was one". Even his political opponents mourned their loss. To this day, many see him as our greatest Prime Minister. The inscription on his gravestone read:

His country was his pride
His brother man was his cause.

So what can we say about Curtin's health and his fitness to lead? He certainly had a drinking problem, though historians disagree as to when he finally stopped drinking. Moreover, before we all knew about the dangers of tobacco, he smoked about 40 cigarettes a day. Just from the smell of cigarettes, his daughter always knew when he was home.

Did poor health affect Curtin's performance? His contemporaries say yes, at least during the last year of his life. Did the demands of office affect his health? Yes. Did these demands shorten Curtin's life? Surely.

(GB)

WHAT KILLED FRANKLIN ROOSEVELT?

In 1999, the *Sydney Morning Herald* ran a quiz in which readers had to elect a world leader. Candidate A was said to have crooked friends and to consult astrologers; he drank eight martinis a day, smoked heavily and had two mistresses. Candidate B was a decorated war hero, a vegetarian and non-smoker. If you went for B, you'd just elected Adolf Hitler. A would have got you Franklin D. Roosevelt.

Born into a patrician American family in 1882, young Franklin enjoyed many advantages, but good health was not one of them. He had frequent colds, pneumonia, typhoid, appendicitis and tonsillitis. Then in 1921, at the age of 39, he caught polio, which was misdiagnosed at first. It is said that for some time he could not even hold a pen. Despite all his treatments and exercises, Roosevelt could never again walk unaided. He could only get about in a wheelchair, or else by leaning on someone's arm while using a cane and wearing leg braces. Roosevelt himself and the media treated his disability largely by denying it. Press photos never showed his wheelchair.

He went on to become Governor of New York in 1928 and President of the United States of America during the Great Depression. Roosevelt started his third term in 1941 at the age of 58. In December 1941, the Japanese attack on Pearl Harbor brought the USA into the Second World War. Later, Roosevelt, Winston Churchill and Joseph Stalin, the "Big Three" allied leaders, met at Tehran and at Yalta.

Admiral Ross McIntire, an ear, nose and throat specialist, treated (perhaps overtreated) Roosevelt for his chronic sinus infection. But there were other problems, too: by 1943, McIntire was secretly limiting the altitudes at which the president could fly. The next year, Roosevelt would occasionally nod off while talking. Once he even blacked out while signing a letter. Accordingly, in March 1944, he had a hospital check-up. His blood pressure was high, he was blue, breathless on movement and kept coughing up yellow phlegm. In brief, he had hypertension, a heart that was failing to pump blood effectively and chronic bronchitis or obstructive lung disease.

Cardiologist Dr Howard Bruenn was shocked, but Dr McIntire, as his senior, put out yet another reassuring public statement. A medical panel recommended digitalis to help the heart to pump, a reducing diet, less smoking and more rest. Roosevelt followed this advice and for the last year of his life, Dr Bruenn stayed constantly at his side. But reportedly no-one told him how ill he was; nor did he ask. On Inauguration Day, January 1945, as Roosevelt started his fourth term, his son James wrote: "He looked awful, and regardless of what the doctors said, I knew in my heart that his days were numbered." Indeed, his father gave James instructions about his (Franklin's) funeral. Later, James said, "I never have been reconciled to the fact that Father's physicians did not flatly forbid him to run."

It is said that during his last few months of life, Roosevelt's systolic blood pressure sometimes reached 300 (about double the normal level).

Now let's turn to the Yalta Conference of February 1945. Dr Ronald Winton wrote: "There [at Yalta] Roosevelt, Churchill and Stalin — three men of whom all were sick but not all were equally wily — redrew the map of Europe ... "

Supporting Dr Winton's reference to Stalin's wiliness are several accounts stating that Stalin manipulated both Roosevelt and Churchill. How? Stalin induced them to make long arduous journeys to meet him. At Tehran, he kept them waiting a full day before showing up. By placing Roosevelt and Churchill in widely separated residences, Stalin even made it hard for them to meet without him. At dinners, Stalin liked to get foreigners drunk by proposing toast after toast. He kept refilling his own glass with what looked like vodka but really was only iced water.

By now, Roosevelt was very ill. At Yalta, Sir Alexander Cadogan, of the British Foreign Office wrote: "Whenever he was called on to preside over any meeting, he failed to make any attempt to grip it or guide it, and sat generally speechless." Churchill's physician Lord Moran called Roosevelt "a very sick man ... I give him only a few months to live".

Moran was spot on. Doctors attributed Roosevelt's death in April 1945 to a sudden large bleed into his brain, but the absence of an autopsy fuelled speculation. I believe that prominent men should have a postmortem and the findings should be made public.

What did kill Roosevelt? Surely it was not just his diet that caused Roosevelt to lose about 10 kilograms? Colleagues noted his poor appetite and weight loss, while minders censored photos taken of him in 1945. During the 1930s and early 1940s, some photos show a slowly growing mole over Roosevelt's left eye. In 1943, the mole was gone, but a scar was left behind. Had doctors removed a melanoma (a very malignant skin cancer) and kept it very quiet? Some kind of cancer seems likely, especially in a heavy smoker.

It is natural to agree with the critics who argue that Roosevelt was quite unfit for office for his fourth term in 1945: "It was irresponsible and dangerous to allow an invalid, possibly a dying man, to engage in negotiations with Stalin and Churchill at Yalta that would virtually recast the maps of Europe and Asia and influence the lives of millions of people."

(GB)

HORATIO NELSON: HERO AND ADULTERER

Way back in 1966, when I was a poor young Aussie in England, I used to enjoy the advertisement for Cockburn Port in the London tube. It went something like this:

Said Lord Nelson to Lady Emma:
"Though Cockburn's a drink well-known and renowned,
When I try to pronounce it, my tongue runs aground."
Replied Lady Emma, with a twinkle:
"My Lord, it's quite simple,
The O is as long as a midsummer's day,
And you just turn your blind eye to the C and the K."

All these years, I've been wondering about Horatio Nelson (1758–1805), his blind eye and Lady Emma.

We still remember him for his naval victories and his extramarital affair with Emma Hamilton — "the worst-kept secret in the long history of British hypocrisy". Let's look at Nelson's naval career first.

As a boy, Nelson was frail and often ailing; whenever he ventured onto a boat, he even got seasick. But he must have had the right stuff. At 14, on an expedition to the North Pole, he nearly became dinner for a polar bear that he was trying to shoot. At 16, he sailed as a midshipman to the East Indies. There he contracted malaria, which kept recurring. Nelson was only 20 when he became post-captain of the frigate *Hinchinbrook*. Most of his crew died of a fever. Later, while in action against the Spaniards in Nicaragua, he was lucky to survive "Yellow Jack" (perhaps yellow fever).

In 1780, he was invalided home with excruciating pains, paralysis and loss of feeling in his limbs. The next year, Nelson wrote: "... knocked up with scurvy; having, for eight weeks, myself and all the Officers lived upon salt beef..." Sad to say, this was a generation after James Lind's treatise on preventing sailors from getting scurvy. When Nelson married a widow, Frances Nisbet, a guest observed that the groom was "more in need of a nurse than a wife".

In 1794, Nelson was ashore besieging Calvi in Corsica when a shell from a French 18-pounder exploded near him. Hence the legend of his blind right eye. We cannot now be sure just what eye injury Nelson suffered, perhaps a fragment in the eye or a bleed inside the eye. Nelson himself wrote, "The surgeons flatter me that I shall not entirely lose the sight; at present I can distinguish light from dark but no object."

In 1797, during a disastrous assault on the island of Tenerife, a musket ball smashed his right arm. The operation report stated: "Admiral Nelson: Compound fracture of right arm by a musket ball passing a little above the elbow, an artery divided; the arm was immediately amputated." Within two days, he was writing letters with his left hand.

In 1798, Nelson on the *Vanguard* attacked 16 line-of-battle French ships to win the historic Battle of the Nile. This victory cut off Napoleon's army in Egypt and ruined his Egyptian campaign. But a metal fragment cut open Nelson's forehead to the bone, and a flap of scalp poured blood as it fell over his good eye.

At the battle of Copenhagen in 1801, he claimed to have put his telescope to his "blind" eye and hence missed seeing the signal to leave the action. This was the second time Nelson ensured a major victory by openly disobeying orders.

Not long after, he complained of "terrible spasms of heart-stroke which nearly carried me off", and thought that his heart was actually breaking.

Nelson's greatest victory came in 1805 at the Battle of Trafalgar, fought between Cadiz and the Strait of Gibraltar. Here 27 English ships faced a larger Spanish and French fleet.

He insisted on leading his fleet into action. Even worse, he stayed on the quarterdeck of the *Victory* in full view of a French sharpshooter only 15 metres above him in the mizzenmast of *La Redoubtable*. Inevitably, a musket ball struck him.

When he heard that his fleet was winning, the dying Nelson murmured: "Thank God, I have done my duty ... Remember that I leave Lady Hamilton and my daughter to my country." Nothing about the faithful wife he had deserted!

Nelson's victory not only shattered Napoleon's plans to invade Britain but also secured British naval supremacy for over 100 years.

The *Times* wrote: "We know not whether we should mourn or rejoice. The country has gained the most splendid and decisive Victory ... *but the great and gallant Nelson is no more.*"

None mourned more bitterly than Nelson's own sailors: "Chaps that fought like the Devil sit down and cry like a wench."

After lying in state in the Royal Hospital at Greenwich, Nelson was buried in the crypt of St Paul's Cathedral.

Nelson may have been a naval hero, but he outraged English society with his lifelong extramarital affair with Emma Hamilton, the wife of Sir William Hamilton, British ambassador to the Court of Naples.

Daughter of an illiterate blacksmith, Emma was pregnant by the age of 16. Her beauty had gained her a protector who paid for her to learn dancing, singing and acting. Later, this young man, being hard up, sold her to his uncle, Sir William Hamilton, British ambassador to the court of Naples. She so entranced Hamilton, 35 years her senior, that he married her.

After his victory over the French at the Nile in 1798, Horatio Nelson came as a conquering hero to Naples. The Hamiltons welcomed him into their house and Emma welcomed him into her bed. She was 33, he was 40; they remained lovers while both lived. But Emma never left her husband; she, her husband and her lover went everywhere together, almost like Siamese triplets.

People in Naples were relaxed about this, but when the trio returned to England and Nelson did not return to his wife, Fanny, society was outraged. Nor did the English approve when Emma bore Nelson a daughter, Horatia, though Emma tried to pass her off as her goddaughter.

Emma in her prime was a beauty, painted by many artists. Her many friends praised her intelligence, generosity and high spirits, while her many enemies complained of her coarse manners and heavy drinking, eating and gambling.

Hamilton died in the arms of Nelson and Emma. Two years later, Nelson died in the arms of victory at Trafalgar. Both men left Emma generous bequests, but her extravagance and gambling soon left her penniless. In 1813, she went to debtor's prison for a year. Emma's lonely descent into poverty, illness and drink ended when she died in Calais in 1815.

And poor, deserted Fanny? She had the last laugh, outliving both Nelson and Emma.

(GB)

Rasputin: The man who would not die

On 18 December 1916, divers fished a frozen body out of the River Neva in the Russian capital Petrograd (St Petersburg). The death of Grigori Yefimovich Rasputin was as violent and brutal as his life.

He was born circa 1869. At 18, in a monastery he met the Khlysty (Flagellants) sect. He perverted their beliefs into his own teaching that one was nearest to God when one felt "holy passionlessness". One could best reach this state through debauchery and sexual exhaustion. So to be saved, you had to sin! Preaching that contact with him had a purifying, healing effect, Rasputin seduced many eminent women.

He married but did not settle down. Instead, he wandered to Greece and Jerusalem, living off donations from peasants and promoting himself as a holy man, prophet and healer.

Rasputin was strongly built, with wide shoulders and long arms. His unkempt beard reached to his chest. Above all, it was his piercing blue-grey eyes that commanded attention. His enemies could not stand his atrocious table manners, body odour or illiteracy, but when he reached the Russian capital in 1903, he made a big impression. Tsar Nicholas II noted breathlessly in his diary: "We have got to know a man of God, Grigori."

Nicholas's wife, Tsarina Alexandra, granddaughter of Queen Victoria, carried the gene for haemophilia (a hereditary disorder in which blood does not clot adequately), and her fifth child and only son, Alexis, heir to the Russian Empire, had the disease. At that time there was no specific treatment for it. His parents had tried both orthodox and alternate healers, with no luck.

In some unknown way, Rasputin could relieve the boy's agonising, terrifying attacks of internal bleeding. He did this not just once but many times, over at least eight years. He soon became indispensable to Alexandra and Nicholas. Rasputin dominated Alexandra who dominated Tsar Nicholas. Rasputin even advised the Tsar on ministerial appointments. Anyone who spoke against him risked dismissal. The hatred that most of the capital felt for Rasputin dragged the Romanov dynasty into disrepute. In time, his very name became a byword for corruption, evil and debauchery, and he survived an assassinationn attempt in 1914.

During the First World War, when Nicholas went to the front to command his troops, the unpopular Alexandra took over political command in the capital. Many people assumed the worst: that the despised Rasputin was the power behind the throne and also a German spy, while Alexandra was his lover. By November 1916, the Russian Empire was on the brink of collapse from the devastation of war, rampant inflation and political instability.

In late 1916, the young Prince Youssoupov enlisted a doctor and three other conspirators (including two relatives of the Romanovs) in his plan to murder Rasputin.

He invited Rasputin to visit him at midnight in a basement room of Youssoupov's palace to meet a beautiful lady. It is said that Dr Stanislaus Lazovert ground up enough poison (thought to be cyanide) to kill a dozen men and mixed it inside a chocolate cake.

In a desperate attempt to be "saved", Rasputin had to subject himself to repeated acts of sin!

He also laced some wine with poison. Even before he arrived, Rasputin reportedly had already drunk at least a dozen bottles of Madeira. Though Youssoupov fed him poisoned cake and wine, Rasputin did not turn a hair. At last the panicky prince shot his guest in the chest with a Browning pistol.

Rasputin collapsed; blood was oozing across his white silk shirt; Dr Lazovert confirmed the death. At three in the morning, the conspirators went upstairs to celebrate.

But something made Youssoupov return to the basement and check the body; he found no pulse. But then Youssoupov was terrified to see Rasputin's eyes twitch, then open ("the green eyes of a viper — staring at me with an expression of diabolical hatred"). With a wild roar, Rasputin rose; Youssoupov fled with the "corpse" following. Finally, another conspirator shot Rasputin in the back and the head while Youssoupov beat him around the head with a club.

This time they tied up Rasputin, before driving to the river, where they pushed him through a hole in the ice. But in their haste they forgot to attach the weights they had brought along.

In less than 24 hours, all of Petrograd celebrated Rasputin's death. Crowds sang, men embraced each other and champagne flowed. The *Times* of London called Youssoupov "The Saviour of Russia". But censorship prevented the Russian press actually mentioning names. One intriguing account started: "A certain person visited another person with some other persons. After the first person vanished ... "

The body was found 200 metres downstream and retrieved with grappling hooks. Surprisingly, it is said that the autopsy showed that Rasputin had not been dead when he was dumped in the river, but had drowned there. Hearing of the autopsy in progress, Alexandra had it halted at once. But no-one has found the report of any autopsy. Hence many doubts remain; for example, had the conspirators castrated him, as it has been claimed?

By killing Rasputin, the conspirators hoped to destroy the pro-German influence at court and to save Russia. But nothing changed; the incompetent ministers stayed on, as did the Romanov policy of repression. In a prophetic letter to Nicholas, Rasputin himself predicted that if nobles or Nicholas's relatives killed him (Rasputin), "none of your children or relations will remain alive. They will be killed by the Russian people."

Within three months, bloody revolution swept away the whole regime. The Bolsheviks came into power, dug up Rasputin's body and burned it.

(GB)

PIZARRO: THE MISPLACED CONQUISTADOR

In the name of Christ, he destroyed a fruitful empire, bringing nothing but disaster ... He represents the dark side of man — Man the Destroyer.

HAMMOND INNES, THE CONQUISTADORS

Born in Spain about 1478, young Francisco Pizarro was illegitimate, illiterate and neglected by his poor parents. When some of the pigs he was tending escaped, Pizarro also ran away.

After fighting with the army in Italy, he made his way to the New World and became chief lieutenant of the explorer-soldier Vasco de Balboa. In 1513, they marched across the Isthmus of Panama and were the first Europeans to reach the Pacific Ocean. This discovery opened the way to new conquests. The smallpox that the Spanish brought to Panama also helped to overcome indigenous resistance.

Soon Pizarro became a partner of another soldier, Diego de Almagro, and Fernando de Luque, a priest, agreeing to share all spoils equally. They could not resist rumours of the rich southern kingdom of the Incas. On their first two expeditions during the 1520s, supplies ran out, many men died in the jungle and Indians attacked them. Pizarro himself suffered seven arrow wounds.

After crossing what is now Ecuador, they explored to nine degrees south but finally had to turn back. Pizarro returned to Spain to ask the Emperor Charles V to authorise another mission. In return for Pizarro's promises of treasure, the Emperor named him governor of the province of New Castile along the west coast for almost 1000 km south of Panama.

In 1531, Pizarro set out again from Panama with 180 men and 37 horses. Disease and skirmishes plagued them. It took nearly two years for the survivors to enter present-day Peru and make contact with the Inca emperor. The emperor Atahualpa's army of over 30 000 was camped outside the town of Cajamarca. Messengers told him of men with strange clothes, beards and unknown animals advancing through the deep valleys of the Andes.

Did Atahualpa believe he was immortal? Or did he see no danger from the tiny band of Spaniards? At any rate, it is said that he and 6000 soldiers came unarmed to meet them.

A story goes that a Spanish priest held out a Bible and asked the Emperor to accept Christianity and the authority of Emperor Charles V. Atahualpa, who in Inca culture was himself divine, flung down the Bible. At Pizarro's signal, the hidden cavalry charged. They killed thousands and took many prisoners. Pizarro offered to spare Atahualpa in return for a large hall filled with gold. Though the Indians collected much of the ransom, Pizarro broke his promise. Rather than be burned alive as a heathen, Atahualpa chose to be baptised and then garrotted.

For eight years, Pizarro ruled the greatest of all the Spanish American provinces. But, wrote Hammond Innes, "the seeds of disaster were ingrained in his own nature". It was a rule of terror, murder, rape and plunder.

Pizarro's partner Diego de Almagro conquered much of Chile, but finding little booty there, returned to Peru. The two fell out and in 1538 Pizarro had his partner tried and executed. Inevitably, Almagro's son sought to avenge the death of his father. On Sunday 26 June 1541, Pizarro was dining with about 20 supporters. When he heard a tumult outside, he ordered the front door locked, but the conspirators broke through. Were there only seven, or as many as 25? Accounts disagree.

Many guests fled. Pizarro himself did not even have time to put on his armour, but he killed two men. After he stabbed a third man, the conspirator standing behind this man shoved him forward, impaling the dying man on Pizarro's sword. While Pizarro tried to pull his weapon free, he received a disabling rapier blow to the throat. All the conspirators dipped their swords in the tyrant's blood. Pizarro died, as he had lived, by the sword.

Carlos Fuentes in his *Republics of the Indies*, quotes Roman y Zamora, who calls Pizarro's soldiers "the worst men who ever set out from any nation and who ... brought the greatest dishonour to the kings of Spain". He and his conquistadors had wiped out the whole Inca nobility and destroyed Inca culture.

Innes Hammond wrote: "Despite his base qualities, Francisco Pizarro remains a perversely heroic figure." Indeed, over the centuries, he has had many admirers, even worshippers.

At first Pizarro was buried behind the cathedral in Lima, but his remains were as restless as the man himself. They were moved twice in the sixteenth century. In 1606, when the cathedral was rebuilt, the remains went into the new church. Earthquake damage later led to other moves.

1891 was the 350th anniversary of Pizarro's death. The mummified body was blessed and reburied in a beautiful stone coffin. Since then hundreds of thousands have come to pay homage; some knelt to pray there.

But in 1977, workmen cleaning the crypt behind the altar found a recess with two wooden boxes filled with human bones and a lead casket on the lid of which was inscribed: "Here is the skull of the Marquis Don Francisco Pizarro who discovered and won Peru and placed it under the crown of Castile." Authorities called in an eminent Peruvian historian, two radiologists and an anthropologist. These experts agreed that the newly found skull was that of Pizarro, but many people kept their money on the mummy.

In 1984, American forensic anthropologist Professor William Marples matched the skull, which had few teeth left, with a skull-less skeleton from the second box. The reunited skeleton was that of a white male aged at least 60. The bones had traces of terrible wounds: four sword thrusts to the neck. Another thrust must have nearly split open the spinal cord. A rapier or dagger had passed through the neck up into the brain. Another thrust had passed through the left eye socket. There were also other wounds: to the arms and hands (probably defensive wounds). All up, there were 11–14 stab wounds that had marked the bones. Clearly, this man had suffered a dreadful, violent death.

Green stains on the heelbones also matched reports that Pizarro had been buried with a Moorish spur. The green was probably verdigris (staining from copper in the spur). Dr Maples was sure that these were indeed the remains of Pizarro.

But if that were so, who was the mummy in the stone coffin? Thorough examination of every bit of remaining skin showed no wounds at all. Even under magnification, the bones showed no fractures, chips, scratches or cuts. This man had led a quiet life and died a quiet death. He may have been a scholar or churchman, but certainly not a conquistador.

Now Pizarro's bones rest in a chapel of Lima's cathedral. Francisco Pizarro has finally displaced the impostor.

(GB)

President Grant: From West Point to the White House

Ulysses S. Grant won fame as the most successful Yankee general during the American Civil War. Later, he became the first West Point graduate to make it to the White House.

He was born in Ohio in 1822. As a boy he enjoyed farming and riding horses. Later, at West Point, he was a superb horseman, so military intelligence drafted him into the infantry. Grant got engaged to be married, but the threat of war with Mexico delayed the wedding. He distinguished himself at the capture of Mexico City. By 1854, he was a captain, but still could not support a family on his army pay, so he left. Or did his drinking and an unsympathetic commanding officer force him out of the army? In any case, Grant resigned and settled in St Louis.

During the next six years, Grant went from one failure to another. He tried farming (crop prices were low), real estate (he could not collect rents) and clerking, first in a customs house and then in a store. In vain, he stood for county engineer; sometimes he resorted to peddling wood.

In 1861, when the Civil War began, Grant was 38 years old; he had already freed his one slave and was happy to re-enlist. His victory at Vicksburg on the Mississippi divided the Confederate (southern) states. Finally, Lincoln made Grant supreme commander of Union forces. On 9 April 1865, at Appomattox, Grant accepted General Lee's surrender. He released Lee and his soldiers on their honour and let the men keep their horses "for the spring plowing". His victory made him a hero in the North, and even Southerners appreciated his generous terms.

In 1868, the Republicans drafted him for the presidential election, which Grant won decisively. He started well, admitting his lack of political experience: "The office has come to me unsought; I commence its duties untrammelled." But he appointed too many personal friends and relatives. Though no-one questioned Grant's own integrity, scandals and frauds implicated his protégés. After his military successes, Grant's presidency was an anticlimax. He left office in 1877, to travel with family in Europe and the Far East. But during a cruise, he lost his dentures overboard and had to give up public speaking.

Back home, Grant retired with savings of about $100 000, which he invested in the banking firm of Grant & Ward. His son, a partner, promised Grant that Ward was a financial genius, but

Ward turned out to be a swindler and the bank failed. To avoid the poorhouse, Grant started writing magazine articles about his wartime experiences.

In June 1884, while biting into a peach, he suffered severe throat and facial pain, which soon became chronic. Finally, a surgeon saw a growth at the base of Grant's tongue and advised Grant to see his own doctor (Barker) at once. But since Barker was in Europe, the stoical Grant did nothing and so lost three months. By October, Grant had an enlarged gland under his jaw. Barker sent Grant to John Douglas, an ear, nose and throat man. A biopsy showed cancer of the right tonsil. Plastic surgeon George Shrady advised against surgery: a controversial decision.

Local treatment included iodoform, gargles of salt water, dilute carbolic acid, and yeast with permanganate of potash. Twice a day, Grant (to save money) took the streetcar to Douglas's office where the latter sprayed his throat with cocaine. At times, Grant also had injections of morphine, cocaine, and reportedly even brandy! Though Grant's drinking and smoking had probably contributed to his cancer, Shrady did allow him the comfort of a few cigars.

Shrady, who also edited the *Medical Record*, published weekly bulletins on the general's condition. Grant's weight dropped from nearly 200 to 146 pounds. After a night of threatened suffocation in March 1855, he spent his days and nights sitting up with his feet resting on a chair.

Once Grant lost his voice, he had to write scrappy notes to his family, friends and doctors.

To provide for his family, Grant pushed himself to write or dictate his memoirs, with his friend Mark Twain as publisher. Twain took the risk of guaranteeing royalties on an unwritten work by a man who had never written a book before and was dying of an agonising, incurable illness.

On 10 July 1885 Grant predicted that in two weeks, his "work would be done". On 23 July with the work off to the printer, he died, at the age of 63. The memoirs were a huge success: "one of the great classics of military literature". In the decade after Grant died, his book sold 312 000 copies.

In New York city, Grant's body lies in a tomb in the General Grant National Memorial.

(GB)

Chapter 7

DEATH

DEATHS OF THE FAMOUS

In the midst of life, we are in death.

THE BOOK OF COMMON PRAYER

Mary, Queen of Scots

If Mary Stuart, Queen of Scots, were living now, she would surely feature every week in our women's magazines. Married three times; a rallying point for the Catholic cause; deposed from her throne; imprisoned for two decades by her cousin Queen Elizabeth I of England. Finally, in October 1586, at Fotheringay Castle, the English convicted Mary of treason.

In February 1587, 300 knights and gentlemen watched her enter the hall, tall and majestic, in a black robe and white veil, with a gold crucifix around her neck.

Mary's crowning glory was her head of thick red hair. She sat calmly on the scaffold while Lord Shrewsbury proclaimed her death sentence. Mary snubbed the prayers offered by the Anglican dean of Peterborough: "I am a

Catholic and shall die a Catholic. Your prayers will avail me little."
After she finished her own prayers, attendants helped Mary to
disrobe and reveal crimson undergarments.

She knelt and, smiling at her ladies, wished them "*Adieu!*
Au revoir."

The first blow missed her neck and cut into the back of her
head. She did not flinch, but only whispered "Sweet Jesus". The
second stroke severed her neck except for one tendon, which the
executioner had to saw through with his axe. As he grasped her red
hair to hold up her severed head, Mary's head separated from her
wig and bounced onto the floor. Onlookers saw a bald, wrinkled old
woman; some claimed that her lips moved for another 15 minutes.

Her small Skye terrier scurried from under her robe and lay
down "betwixt her head and body, and being besmeared with
blood, was caused to be washed".

Emperor Charles V

The Holy Roman Emperor Charles V (1500–1558) wanted to make
sure that officials ran his funeral just as he wanted it. So he kept
rehearsing it, again and again.

He ordered his tomb be erected in a monastery chapel and had a
procession of his servants, each holding a black taper. Following
them came Charles himself dressed in a shroud. As the hymns
struck up, he would lie in the coffin, joining in the prayers for his
immortal soul. As he wept, the priests would sprinkle holy water
over him. After the congregation filed out, Charles would rise from
his coffin and head back to his bed.

Jeremy Bentham

Philosopher Jeremy Bentham (1748–1832) set out the principle of
Utilitarianism: "the greatest good for the greatest number".

Naturally he encouraged people to donate their bodies for dissection and so advance medicine. When he himself died, select people got an invitation:

> Sir
>
> *It was the earnest desire of the late JEREMY BENTHAM that his Body should be appropriated to an illustration of the Structure and Functions of the Human Frame ... the honour of your presence ... is requested ...*

It was a dramatic dissection, with a violent thunderstorm shaking the building. The surgeon, his face "as white as that of the dead philosopher before him" had to operate between the flashes of lightning.

In his will, Bentham instructed the surgeon to reassemble his skeleton, wire together the joints and pad out the whole with straw and hay. But instead of keeping Bentham's head, he fitted a wax model. The end result, Bentham's auto-icon, still sits grandly behind glass in University College, London. Bentham's actual head sits between his feet!

Franklin D. Roosevelt

American president Franklin D. Roosevelt (1882–1945) wanted a simple funeral. In 1937, he wrote four pages of instructions: he should not lie in state; no hearse; a simple dark wood coffin; no embalming; his grave not to be lined with cement, bricks or stones. Clear enough?

So he *was* embalmed, *did* have a copper coffin, *did* go in a Cadillac hearse to a vault that *was* cement-lined. Why? He had left his orders in a private safe, which no-one found until after his funeral.

William Shakespeare

We remember some famous people by verse. That on Shakespeare's tomb reads:

> Good friend, for Jesus' sake, forbear
> To dig the dust enclosed here:
> Blest be the man that spares these stones,
> And curst be he that moves my bones.

Prince Frederick Louis

Then there was Prince Frederick Louis (1701–1751), son of King George II and father of George III. Fred was an intriguer whom even his parents could not stand. This anonymous verse expresses the English dislike of not just Fred but the whole brood of Hanoverians:

> Here lies Fred,
> Who was alive and is dead;
> Had it been his father,
> I had much rather;
> Had it been his brother,
> Still better than another;
> Had it been his sister,
> No one would have missed her;
> Had it been the whole generation,
> Still better for the nation;
> But since 'tis only Fred,
> Who was alive and is dead,
> There's no more to be said.

A few last words

James Rodgers, American criminal, shot in 1960, when offered a last request, asked for a bulletproof vest.

General Sedgwick, American Civil War commander, peering over a parapet: "They couldn't hit an elephant at this dist ... "

Dr William Palmer, Victorian poisoner, hanged in 1856, as he stepped onto the gallows: "Are you sure it's safe?"

Finally, Ned Kelly, hanged in 1880: "Such is life."

(GB)

What really killed Socrates?

Even those Western countries that still have capital punishment reserve it for the most serious crimes. We do not expect civilised nations to execute citizens for their political ideas.

But in 399 BC, the city-state of Athens, renowned as the cradle of democracy, put to death the 70-year-old Socrates. He was the first of the three great Greeks (the others being Plato and Aristotle) who laid the philosophical foundations of Western culture.

Socrates's crimes? His scepticism, his beliefs and his teachings.

He was short, stout and ugly with thick lips, bulging eyes and a flat nose. He may have trained as a stonemason, but others say he had no trade or occupation. Marrying late in life, he had three sons.

Socrates served Athens bravely in the Peloponnesian War against Sparta. When in Athens he spent most of his time in the streets and markets discussing philosophy and ethics with anyone who would listen. Many were aristocratic young men. He taught that every man must "care for his soul". Knowledge of what one ought to do, he said, was a step towards virtue. Conversely, ignorance led to evil. Socrates did not promote any one philosophy or creed, but encouraged his listeners to learn to know themselves. Himself feigning ignorance, he would keep questioning them. Thus he exposed their ignorance and so encouraged them to seek the truth that lay within them.

Rulers, he suggested, should be those who can rule well; not necessarily those who can win elections.

In 404 BC, after 27 years, Athens finally lost the Peloponnesian War; the greatness of Athens was gone. "Her treasury empty, farms and olive trees burned, navy destroyed, trade ruined, two-thirds of her citizen body dead from disease or war" wrote Dr Gordon Daugherty.

The crushing defeat brought repression, terror and the settling of old scores. The charges that the authorities now brought against Socrates were very general: not believing in the gods of the city, introducing new gods, and corrupting the political morals of youth.

The jury voted against him, but the vote was close. They asked Socrates what his punishment should be. He claimed to be Athens' gadfly of truth, rousing its citizens from their stupidity and greed. Hence, as a public benefactor, he asked the state for free dinners and lodging for the rest of his life!

Had he shown humility, the jury might have simply exiled him. As it was, they condemned him to die. Even in prison, they let him see visitors. His close friend Crito bribed the guards to let him escape, but he refused, since he believed that citizens should obey even unjust laws.

What do we know of Socrates and his death?

Since he himself wrote nothing, we depend on the evidence of others. We have a famous painting of the death scene, but that dates from 1787. Plato tells us of Socrates holding a long dialogue on death with 14 of his disciples. Towards sunset, the jailer brought the famous cup of hemlock. Socrates asked the gods to prosper his journey to the other world and "raising the cup to his lips, quite readily and cheerfully drank off the poison".

Presently he lay on his back. E. Hamilton and H. Cairns wrote in *Plato: The Collected Dialogues*, "The man [gaoler] after a while ... pinched his foot hard and asked if he could feel it. Socrates said no ... he was getting cold and numb."

He died peacefully. But what was actually in the cup? What did kill Socrates?

By hemlock, we usually mean the plant *Conium maculatum*, also called "poison hemlock" or "spotted hemlock". People sometimes mistake its leaves for parsnip, parsley or celery. It looks a little like a carrot but has a white root.

Most accounts describe hemlock poisoning as being nothing like the peaceful death that Plato described.

Nicander of Colophon (204–135 BC) wrote of the "noxious draft" which made the eyes roll, men totter and crawl on their hands: "a terrible choking blocks the lower throat and ... windpipe; the victim draws breath like one swooning, and his spirit beholds Hades". A later authority described an equally frightening death: "convulsions, coma, violent delirium, salivation and involuntary discharges from the bladder and bowels". From the twentieth century, we also have experiments on animals that match these descriptions of hemlock poisoning.

Did Plato, who idolised Socrates, simply omit undignified details? Perhaps, but in 1995, Dr Gordon Daugherty raised other possibilities. An extra large dose of hemlock could kill faster and with less distress. Might not Crito have bribed the poisoner to do this? Or the cup may have had hemlock laced with opium and even alcohol!

Whatever the exact cause of his death, even today, over two thousand years later, we still honour Socrates.

(GB)

THE MYSTERIOUS DEATH OF EDGAR ALLAN POE

Take thy beak from out my heart,
And take thy form from off my door!
Quoth the Raven,"Nevermore".

POE, *THE RAVEN*

Edgar Allan Poe (1809–1849) was a master of writing mysterious and dark tales to embellish already morbid and macabre subjects. *The Raven*, a snippet of which is quoted above, is a compelling and chilling example. He led a dissolute, shadowy life, yet, disturbed though some of his works appear, Poe wrote with such felicity that he seems the embodiment of the cliche "genius is akin to madness". He died in dramatic fashion at the age of 39 in 1849 when he was found alone and dishevelled in the street following what has always been presumed to have been an alcoholic blowout.

But was it? Reappraising his clinical notes, Michael Benitez, professor of medicine at the University of Maryland, has recently thrown new light on the last days of this fascinating man, concluding that his death was even more extraordinary than his life.

Poe was born in 1809 in Boston, USA, the son of an English actress and an alcoholic American actor. His mother died when he was two, and his sister died some years later in an institute for what was then called "mental defectives".

John Allan, his godfather, brought him up in what was an unhappy, loveless household. Despite that, he received a classical education in Greek, Latin and French, but socially young Poe was a failure. As a youth he ran up gambling debts which his guardian refused to pay, whereupon he joined the army as a cadet at West Point, but contrived to be expelled by the simple expedient of missing parades for a week. By this time he had begun to write and proceeded to marry his 13-year-old cousin.

But he continued to be a troubled soul who attempted to relieve his bouts of depression with alcohol. Yet he became a well-regarded author despite often being seen drunk in public. He became dissolute and emaciated, but personable and engaging enough that, when his wife died of tuberculosis, several women promised marriage if he would stop drinking. He refused all offers, but they inspired him to write some touching essays and poems.

Several editorial posts followed, but chances were squandered when he disappeared for two or three months to pursue his opium and alcohol addiction. A wild spree in Philadelphia early in 1849 was followed by a happy, drug-free summer in Richmond, Virginia, in the arms of a widow, Mrs Sheldon. However, in September he suffered forebodings of death, so departed via the railway to return to Philadelphia but detrained in Baltimore and disappeared, seemingly into thin air. None of his friends ever saw him alive again.

So what happened? Was it alcoholic poisoning or something else that killed him? The answer theorised by Professor Benitez makes for good medical history detective work.

The sequence of events was that Poe resurfaced at 7 a.m. on 3 October five days after leaving Richmond, when he was found unconscious and in an untidy and rumpled state lying across two barrels under the steps of the Baltimore Museum. On admission to the Washington College Hospital there was no sign of injury or smell of alcohol. This lack of evidence of alcohol has been disputed, but Benitez claims it is in the hospital notes.

In any event, Poe died four days later, but in that brief period we now know there were produced a set of symptoms so unusual as to throw doubt on the usually accepted diagnosis of a drug-induced death.

Initially, the unconscious patient was unresponsive, but at 3 a.m. on 5 October he developed an irregular and thread-like pulse, sweating, tremulousness and visual hallucinations. He remained thus for 28 hours when he abruptly became tranquil, alert and orientated.

He was offered alcohol as a stimulant, but vehemently refused to drink anything, water included.

On the third day, clouding of his mental state returned. By late that evening he was delirious and so combative he had to be restrained. He drank water only with the greatest difficulty and under vigorous protest, foaming at the mouth in the attempt. He remained thus until he died the following day, 7 October 1849, shortly after he had murmured, "Lord, help my poor soul." He was buried with little ado in Baltimore's Presbyterian Cemetery.

What was the cause of death? Benitez, after reviewing the various causes of delirium such as endocrine, vascular, neoplasia (tumour), central nervous system (CNS) infections, metabolic, nutritional and toxic, thinks he has the answer, and could well be right.

Knowing the alcoholic background, he dismisses first "a nutritional cause" as in, for example, Wernicke's syndrome, a brain degeneration consequent on long-term alcohol ingestion (because there were no ocular motor signs and the condition was progressive, not relapsing as in Wernicke's) and Pellagra, the vitamin deficiency malady (there were no diagnostic dermatological signs and again the condition was progressive).

Second, "a toxic cause" through alcohol poisoning is discounted because we know Poe had been on the wagon for six months and alcohol was not in evidence on admission. Further, and crucial to the diagnosis, he became acutely ill, briefly recovered, then relapsed, which is not the story with alcohol.

Again opiate withdrawal gives a different story, including "goose flesh" or "cold turkey", and is not recorded here.

Benitez critically looked at and dismissed "relapsing fever", yellow fever, malaria, epilepsy, eastern equine encephalitis and several other rare pathologies as the story did not fit. However, in his view one viral brain infection did fit — rabies encephalitis.

This condition is marked by the acute onset of confusion, hallucinations, combativeness, sweating, salivation and spasms. These presentations are episodic and between times the victim is calm and lucid. The patient refuses to drink, indeed is terrified of any fluids, as they produce intense and involuntary laryngeal spasms resulting in frothing at the mouth. All this happened to Poe.

The malady, of course, is contracted via the saliva of an infected animal, yet Poe gave no such history. The basic cause was to be discovered by Pasteur, about 40 years later, so was ill understood in Poe's time and the bacteriology not at all.

Actually, it seems that the lack of animal contact is no bar to the diagnosis as fewer than a third of cases can recall such a contact, mainly due to the fact that the virus tracks along the nerves to the CNS very slowly (up to a year from the foot) so the link is forgotten. No vaccine for rabies existed in the 1840s, and if the brain was reached, then, as now, the condition was 100% fatal.

Benitez proposes that Edgar Allan Poe died of rabies after an initial but long-forgotten exposure, possibly from a cat, an animal of which Poe was inordinately fond.

Charles Baudelaire, the French poet, on hearing of Poe's death, somewhat portentously commented it was "almost a suicide, a suicide prepared for a long time". But, of course, he was relating it to the commonly held belief that the American had died of alcohol poisoning.

Were alcohol and/or other drugs responsible or was it rabies? We shall never know for certain, but perhaps it was prophetic that the imaginative but flawed genius had earlier written in *Ulalume*:

> *The skies they were ashen and sober;*
> *The leaves they were crisped and sere —*
> *The leaves they were withering and sere;*
> *It was night in the lonesome October*
> *Of my most immemorial year.*

(JL)

Chapter 8

Diseases

Iodine: Protecting our children's intelligence

Iodine deficiency has condemned millions of children to cretinism (mental and physical retardation), tens of millions to mental retardation and hundreds of millions to subtle degrees of mental and physical impairment.

JAMES GRANT, FORMER DIRECTOR OF UNICEF

Iodine deficiency is so easy to prevent that it is a crime to let a single child be born mentally handicapped for that reason.

WORLD HEALTH ORGANISATION, 1978

The bad news in Australia is that, unless we act now to prevent it, our children and grandchildren may become mentally deficient. We are not talking about infectious diseases like meningitis or vitamin deficiencies like scurvy, beriberi or rickets. We are talking about a simple lack of iodine. One teaspoon of iodine in a whole lifetime is all we need.

But lack of iodine in pregnant women, their unborn babies and in young children is the world's single most important cause of preventible mental retardation. Brain development depends on the thyroid gland which produces an essential hormone that contains iodine. To make enough hormone, the thyroid gland needs enough iodine. Professor Creswell Eastman, consultant to the World Health Organisation [WHO] says: "Very subtle hypothyroidism [caused by a thyroid gland that is not putting out quite enough thyroid hormone] in ... pregnancy can have serious consequences when you study IQ in the offspring." Too many affected children are still being born. Around the world, 1.5 billion people living in iodine-deficient environments are still at risk of brain damage.

When we think of thyroid problems, we think of goitre, which is an enlarged thyroid gland causing swelling of the neck. Goitre is one result of the lack of iodine, but it is far less important than the effect of iodine deficiency on the brain.

The general term iodine deficiency disorders (IDD) covers all effects of iodine deficiency. In early life, these include abortions, stillbirths, birth defects, infant deaths, mental deficiency, deafness and mutism. Childhood forms include mental and physical retardation.

Where does our iodine come from? Iodine occurs naturally in low concentrations in our environment, mainly in soil and water. Many marine plants and animals (for example, marine fish, shellfish and seaweed) have a lot of iodine. Conversely, iodine deficiency occurs especially inland, in hilly or mountainous areas and in flooded river valleys. The world's most iodine-deficient areas include the European Alps, the Himalayas, the Andes in South America and the vast mountains of China.

Good news: for about 18 years, Australian health workers led by Professor Eastman have been working successfully on this problem in China. Artificial iodine supplements come as injections, tablets, drops, or as additives to salt, bread, water, milk, oil, soya sauce or fish sauce. The cheapest and most effective supplement for large populations is iodised salt. Chinese authorities are working towards universal salt iodisation. Their salt industry is being modernised so it can produce 8 000 000 tons of iodised salt each year!

A survey in 1997 showed that people in all provinces of China except the autonomous region of Tibet had reached the target levels of body iodine. But in Tibet iodine-deficiency disorders are the major public health problem. Over half the people suffer from goitre and at least 30% of Tibetan children are seriously impaired intellectually. In the most iodine-deficient areas, the average IQ of children is only 85. Professor Eastman calls this "a disaster".

The Elimination of IDD in Tibet Project, in which his team is also playing a large part, includes the building of two plants to produce iodised salt for all Tibetans and their livestock. At the same time, all infants and about one million women of childbearing age (i.e. those most vulnerable to iodine deficiency) will get iodised oil once a year. AusAID is supplying $2 million for this iodised oil.

What is our own situation in Australia? More good news: in a well-off country like Australia, we should be able to prevent iodine deficiency. For some decades we have believed that lack of iodine occurs only in Tasmania. This is why in other states we have not regularly monitored people's iodine levels.

When he is wearing his Australian hat, Professor Eastman runs the Australian Centre for Control of Iodine Deficiency Disorders (ACCIDD) at Westmead Hospital in Sydney. Over the last two decades, the centre has occasionally surveyed the iodine content of urine from small samples of Australians. But it's not all good news: in the last eight years, residents of Sydney have shown a gradual fall in these levels. Other small surveys of pregnant women in Sydney and of people in Tasmania also show low iodine levels.

Why are we getting short of iodine? Firstly, for over 30 years, we drank milk that had been in contact with iodine-containing solutions used to clean milk vats, but new cleaning solutions with less iodine are now coming into use. Secondly, we are using less iodine-enriched salt than we used to.

What to do? Professor Eastman recommends an Australia-wide survey to confirm the low iodine levels found so far. He also advises publicity to stress the dangers of iodine deficiency, especially in pregnancy and early childhood. He advises all pregnant and breastfeeding women to use iodised salt and also take some other form of iodine.

But above all, he is pressing for Australia to legislate for universal iodisation of salt. He wants iodine added to all salt eaten by humans and by animals. This recommendation has the strong backing of WHO. Moreover, around the world, 118 countries are now doing this. Professor Eastman can see no good reason why Australia should be among the exceptions.

The cost? Less than 10 cents per person per year. The pay-off? Protecting our children and grandchildren.

(GB)

THE GENESIS OF LEGIONNAIRE'S DISEASE

Between 21 and 24 July 1976, the 58th annual convention of the Pennsylvania section of the American Legion was held in Philadelphia (City of Brotherly Love, no less). The headquarters for this jovial meeting of ex-servicemen was located in what I shall tactfully call Hotel A.

At the same time, the 56th annual get-together of the local American Legion Auxiliary was being held in another hotel, Hotel B. Both affairs included family members, and there were about 1850 delegates altogether. Besides the usual speechifying sessions, there was a parade, a testimonial dinner, a dance, caucus meetings for the king makers and a breakfast. Obviously, a time of festivity and back-slapping bonhomie: no-one was to know of the approaching cataclysm.

Besides the formalities, between times there could often be seen unofficial huddles of delegates in the lobby of Hotel A, or on the pavement outside for the less well-favoured. Moreover, for those with more voting clout, suitable refreshments were served in one or other of the 13 hospitality rooms which had been reserved by the 13 men hopeful of office. District branches and past office-bearers had similar rooms scattered throughout several hotels and in all of these beer, whisky and simple snacks were served to the faithful.

All this must sound familiar to any regular conference-goer. But on this occasion to this time-honoured formula there were added one or two other seemingly unimportant details which in the end were to prove, well, fatal.

For instance, the apparently trivial snippet that Hotel A had been built in 1904, though modified and renovated since, and that guests were accommodated in 700 rooms between the second and 16th floor was to become significant when bracketed with the facts that the meetings were held on the first and 16th floor, and that the out-of-date air-conditioning system consisted of two water chillers from which cool water was circulated to only 60 units in a few strategic places in these areas. At the time nobody gave it a thought. Why should they?

Well, this was why: three days after the conference ended, Legionnaire Ray Brennan died of a pneumonia-like illness in Philadelphia. By the next day 117 ex-servicemen who had been at the convention were seriously ill; six were to die.

Other victims with serious symptoms followed, and when, by 10 August, it was clear that no new cases were developing, it emerged that 182 persons associated with the meetings had been stricken. Of these 147 (81%) had had to be hospitalised and 29 (16%) had died.

Typically, symptoms usually began seven days after the start of the four-day conference, and about three days after returning home. They comprised malaise, headaches, a rising fever with rigours and a non-productive cough. Chest pains were experienced in a third of cases and in 90% X-rays showed patchy areas of consolidation within the lungs.

In most the illness progressed over two or three days and then slowly settled, with radiological evidence of improvement lagging behind clinical recovery. The older the victim, the greater the risk of death. This occurred at a median of ten days after onset and was associated with shock, widespread and choking consolidation and eventual collapse. No single antibiotic was clearly effective.

Of the 182 cases, 142 were males and the age range was from three to 82. One hundred and forty-nine had attended as delegates to the Legion Convention, 17 were family members, one was a hotel employee, and the other 32 had attended other conventions immediately after the ex-servicemens' assembly. Eighty-four of the sufferers had actually stayed at Hotel A, and all had visited it.

Thirty-nine other people who had not attended any conference activities but who had walked down Broad Street past Hotel A at about the same time also fell ill with pneumonia. However, although five died, being non-delegates they did not meet the

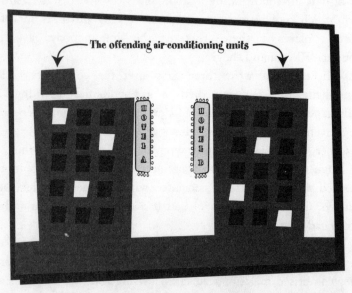

The offending air-conditioning units

The incredibly elusive Legionnaire's Disease had experts studying the Legion Convention Mystery deaths quite stumped. It was later found that the culprits were the air-conditioning units ... they were not that cool after all!

epidemiological criteria of what was now called Legionnaire's Disease, so were said to be suffering from 'Broad Street pneumonia'; a nice distinction probably lost on the grieving relatives.

As epidemics go, this outbreak of Legionnaire's Disease was pretty small beer. Nonetheless, it became an instant Roman holiday for the media and a chastening event for health professionals.

Under the full glare of the public spotlight, the world's best medical brains laboured to identify the causative agent. An intense survey of delegates' movements, hotel staff, roommates (some soul-searching there), hospital admissions of other pneumonia sufferers, all 1002 of Pennsylvania's Legion posts, whether sending delegates or not, weather records, workers in the streets round Hotel A and B (and C and D) was carried out. Nothing, it seems, was left to chance.

Finally, two men who recovered were brought back to "walk through" their congress activities, and, crucially as it turned out, sampling of fluids and air from the physical environment and air-conditioning units of the hotels was undertaken.

From all this there emerged a mountain of information, consequential among which was that a significant number of victims had been in hospitality room A on the 14th floor of Hotel A; in it was located one of the air-conditioning units. No other room showed a difference between control and case numbers.

Further, the mean number of minutes spent in Hotel A's lobby was significantly greater in victims, as indeed it was for those who watched the parade on the pavement outside Hotel A, compared with being a spectator elsewhere. Between case and control there was no significance difference in the food eaten, alcohol (or ice) consumed, or birds, mammals or souvenirs handled. But there was in the water drunk, even though 38% of the sick said they never drank the water, and, of course, none of the Broad Street coterie had done so.

Furthermore, and interestingly, serological tests on the employees of Hotel A suggested the possibility of a low-grade exposure to the malady's causative agent had gone on for several years, thus giving immunity.

In January 1977 it was announced that the pathogen was a previously unknown rod-shaped bacillus. A little later it was admitted that it seemed likely that a number of unresolved epidemics (e.g. in the District of Columbia in 1965, and in Pontiac, Michigan, in 1968) were caused by the same bacillus. Indeed, it was deduced that Legionnaire's Disease had long been with us, but its acquired name now gave it a note of drama which hitherto had been missing.

Eventually, after looking at all the computations, especially noting the implication of loitering in the lobby or the pavement outside Hotel A (over which the exhaust from a cooling unit presumably discharged), or socialising in the suspicious hospitality room, coupled with the significant finding in a special antibody blood test for infections, which for the initiated showed an indirect fluorescent-antibody titre of greater than 1/64 in 29% of the employees, it was concluded that the spread was airborne.

The air-conditioning appeared to be the likely source, though it was thought to be strange that the whole grisly fandango only lasted during an exposure of about two weeks.

Sporadic cases have occurred since, including in hospitals themselves, and, more recently and closer to home, at the Melbourne Aquarium. It would seem that fresh enemies are ever at the ramparts.

(JL)

TUBERCULOSIS CAME HERE WITH THE CONVICTS

A dread disease ... which medicine never cured, wealth warded off, or poverty could boast exemption from — which sometimes moves in giant strides, or sometimes at a tardy, sluggish pace, but slow or quick, is ever sure and certain.

CHARLES DICKENS, *NICHOLAS NICKELBY*

Because of their geographical isolation, Australian Aborigines remained free of tuberculosis (TB) until 1770. But TB caused the first white death in Australia. Fosby Sutherland sailed with Captain James Cook on the *Endeavour* and died of TB (often called consumption) at Botany Bay. Cook buried him on the Kurnell Peninsula, now named Point Sutherland after him. Later Sutherland also became the name of the shire.

Tuberculosis was widespread in England, and the convict transports brought many cases to the colony. Australia's Aborigines, lacking immunity, fell victim not only to TB but also to chickenpox, measles, scarlet fever, whooping cough, diphtheria, smallpox, dysentery and venereal diseases.

What effect, if any, did climate have on people with TB? This was a big question during the first two centuries of white settlement in Australia. Why? Remember that until streptomycin became available in the mid-twentieth century, doctors had no specific treatment for TB. Hence they often advised those who could afford it either to rest up or to travel.

One doctor here wrote that the warm Australian climate made TB progress faster than in Europe. But many English doctors believed just the opposite. Mild climates, whether in Italy or Australia, they insisted, were the best treatment. Indeed, the key English medical journal, the *Lancet*, often carried glowing reports on the good health of our colony. I wonder what made these writers so optimistic about a distant country that most had never seen.

Among hopeful people with TB who made the long trek here were several doctors. Samuel Dougan Bird probably picked up his TB while working at the Brompton Hospital for Consumption in London. First he had six months travel in Europe which did not help. But after only three months in Australia he lost all symptoms and put on six kilos. He was now well enough to set up practice in Melbourne and write a guidebook for other migrants.

But Dr William Thomson who had made six voyages here and also worked in Melbourne insisted that the climate had no effect on TB.

This started, wrote B. Thomas and B. Gandevia in the *Medical Journal of Australia*, "one of the most remarkable medical controversies of the nineteenth century ... it involved not only leaders of the profession and the medical press at either end of the world, but also the lay public and the newspapers".

During the gold rush, migration boomed. Here was a huge reservoir of TB to infect a new generation of Australian-born whites and the even more vulnerable Aborigines. The death rate from TB rose until by the end of the nineteenth century it became the commonest cause of death.

But there were also advances in what we now call public health. Eighteen eighty-two saw the setting up of the NSW Board of Health to control infectious diseases. Soon after came testing of cattle for TB. Authorities opened sanatoria for TB, laid underground sewers and improved water supplies. Next came free medical services for mothers and children; later school medical services. In 1915, the Commonwealth Serum Laboratories opened. Pasteurisation of milk started in the 1930s and TB allowances began in 1944.

Our first Commonwealth Director of Tuberculosis, Dr Harry Wunderly, who himself had TB, persuaded Prime Minister Ben Chifley to fund a national campaign. The 1948 Act increased allowances for people with TB, offered mass chest X-rays and BCG vaccinations and funded TB clinics and hospitals. Also in the 1940s came effective drug treatment, first streptomycin, then PAS and isoniazid.

As the standard of living kept rising, so death rates fell dramatically: from 27 per 100 000 per year in 1948 to 1 in 500 000 per year in 1985.

In 1976, the Commonwealth stopped funding for mass chest X-rays. But then Asian refugees, among whom TB is commoner, started migrating here, and later came HIV/AIDS, which makes patients very vulnerable to TB. The result is that TB in Australia has again become a concern. Younger doctors, because they have not often seen it, may miss it. But the greatest danger is that without strong public support and government funding, measures to control TB will disappear long before TB does.

(GB)

WHEN IS COT DEATH A COVER FOR MURDER?

Five American children died

Waneta and Tim Hoyt were not well off when they married in New York State in 1964, but they did want to start a family.

It was a terrible shock when Waneta found her first child, three-month-old Erik, barely breathing. She could not revive him. Their second child, James, was two years old when he also died suddenly. Julie lived only 48 days. Molly died at three months, as did their last child, Noah.

Waneta adopted a son (who survived) but never forgot her own children. She kept their photos throughout the house and every Memorial Day took flowers to their graves.

Assistant Professor Alfred Steinschneider had cared for the last two Hoyt children. He blamed the deaths on sudden infant death syndrome (SIDS). As well, he taught that all newborn children have brief periods when they stop breathing (apnoea), but that those with apnoea over about 15 seconds are at risk of SIDS.

Moreover, he stated that SIDS had a strong genetic basis; this convinced poor Tim Hoyt to have a vasectomy. Many other bereaved parents wondered if they should do the same.

Steinschneider also believed that home monitoring (with alarms) of the breathing and pulse of infants would reduce the risk of SIDS. He set out these views in a landmark paper which was published in the October 1972 edition of *Paediatrics*.

The effect of this paper was "like lighting a match in a gasoline factory". Anxious American parents flocked by the thousands and spent millions on monitors. To a lesser extent, so did Australians.

But meanwhile, an English researcher, Professor David Southall of North Staffordshire Hospital, was following and monitoring over 9000 babies in a prospective study to confirm the link between brief apnoea and SIDS. In 1982, he startled a SIDS conference by reporting that he could find no such link.

Nor could other workers reproduce Dr Steinschneider's findings. Nurses at his hospital claimed that they had told him long ago that Waneta Hoyt herself was killing her own children. Psychiatrists talk about Munchausen syndrome by proxy: bizarre behaviour when disturbed people injure or even kill their own children as a way of getting attention from doctors and hospitals.

Oddly enough, it was the original paper by Steinschneider that caught the eye of District Attorney William Fitzpatrick. What the paediatrician saw as a deadly genetic affliction, the DA saw as serial killing. In 1994, Waneta Hoyt faced five charges of murder. Steinschneider, now president of Atlanta's SIDS Institute, appeared for the defence. But much of the incriminating evidence actually came from his detailed records on the Hoyt family.

Time magazine asked: "When is Crib Death a Cover for Murder?" Waneta Hoyt confessed to smothering three of her children with pillows, one with a bath towel and one by pressing its face against her shoulder. Later she withdrew these confessions, but she was convicted and finally died in gaol in 1998. She now lies buried in an unmarked grave near the children she killed.

The medical examiner of San Antonio, Texas, said parents like Waneta "usually keep killing until they're caught or run out of children ... Two SIDS deaths [in one family] is improbable, but three is impossible".

Time also cited the case of Marybeth Tinning, whose nine children had died of SIDS and other "natural causes" between 1972 and 1985. "Doctors and friends suspected some rare genetic defect ... even though one of the victims was an adopted son." (*Time*, 11 April 1994.) She was convicted (but not until 1986) of murdering her last child.

Even before the Hoyt trial, two medical experts, Byard and Beal, queried the diagnosis of SIDS in two other American families.

One family had five siblings die and the other had six; at the time, all were attributed to SIDS. Their conclusion: "Munchausen [syndrome] by proxy is worth considering in the differential diagnosis of any case that appears to defy medical logic."

SIDS strikes an Australian family: 1977

On 10 July 1977, Glenn Fitzgerald died of SIDS in Melbourne. He was the eight-month-old fourth child of Kaarene Fitzgerald.

Glenn's death changed Kaarene's life. The pain was not all from the death itself. Many close friends who came to the funeral then faded away. Did these couples fear that they themselves might lose a child? But some others who had been only acquaintances became good friends.

Kaarene was one of many parents who needed to find out what had killed her child. She set up a foundation to raise funds for research into the causes and prevention of SIDS and to provide family support and education for health workers. Since its inception in 1990, the successful "Reducing the risks of SIDS" health promotion program has helped to save the lives of over 3500 Australian children.

Gradually researchers have eliminated many of the original hundreds of possible causes of SIDS. Such research is still continuing, but no single cause has emerged.

Kaarene is sure that serial killers like Waneta Hoyt and Marybeth Tinning would be detected in Australia. The closely knit Australian community of SIDS support groups, researchers and clinicians or the coroner would pick up multiple deaths within one family.

SIDS strikes an Australian family: 1999

Melbourne couple Ryan and Kathryn Bessemer have been together since 1995. On 30 April 1998, their first child, Amy, was born. She was a chubby baby with blond hair who slept through the night from four weeks and smiled early. Her first word was 'Dadda'.

Amy was eight months old when she died of SIDS on 13 January 1999.

After the morning bottle, Kathryn put Amy down. Fifteen minutes later, she found her apparently dead, tried to revive her and called an ambulance, which came within two minutes. Staff continued resuscitation attempts but without success.

Ryan, a computer consultant, was at work when police called him into a side room with the news. They drove him to the Children's Hospital, where a nurse and counsellor waited with him for the ambulance. He still could not believe Amy was dead. Then Ryan had to ring his own mother.

Later, everyone in the family got to hold Amy once more. They bathed her, took photographs, footprints and a lock of her hair. But later still, after the postmortem, they could see only her swollen face and cold hands. Now Amy smelled of formaldehyde. She was no longer the child they had known.

Kathryn felt angry that other members of the family expected her and Ryan to comfort them when they themselves so badly needed comfort. Similarly, they both found it hard to cope with each other's grief while they were battling with their own grief.

Ryan felt suicidal: "The only thing a parent wants, at any cost, is to be reunited with their child." He found the two weeks' leave from work far too short. If he had not had to keep paying off the mortgage on their new house, he would have quit his job. He still goes to monthly SIDS Fathers' Nights.

Eight weeks after Amy died, Ryan went to hear the final postmortem report from the pathologist. Kathryn did not feel up to going. She feels close to her GP, who came to Amy's funeral. Both Kathryn and Ryan welcomed talking to Vivian, a counsellor from the SIDS Foundation.

(GB)

Chapter 9

DISASTERS AND ECCENTRICS

SUFFER THE CHILDREN

*Few doctors practising in the early 1960s .... could read
this book [Suffer the Children] without concluding
'There, but for the grace of God, go I.'*
>   DR FRANCIS ROE, "MEDICINE AND BOOKS",
>   *BRITISH MEDICAL JOURNAL*

In the 1950s, the Distillers group sold Scotch whisky,
vodka, gin and pharmaceutical drugs. If only they had
stuck to spirits.

At Crown Street Women's Hospital in Sydney in 1960
obstetrician Dr William McBride first prescribed a sleeping
tablet/sedative new to Australia: Distaval (thalidomide).
His first patient had severe vomiting of pregnancy; with
thalidomide, she settled quickly. Dr McBride started to
use it not only for morning sickness but also for insomnia
in pregnancy.

Later that year, nurses at Crown Street noted more women than usual with threatened miscarriage. In May 1961, Dr McBride delivered a Mrs Wilson: her baby had a blockage of the bowel and severe defects of both arms. Despite emergency surgery, the baby died after a week. Soon Dr McBride delivered another baby with similar defects; another death. In June came a third. Since the average rate of birth defects was only about 2%, this was a remarkable sequence. All three women had taken thalidomide during their pregnancy.

It was the German company Grunenthal that had first synthethised thalidomide. The *Sunday Times* Insight Team claim that Grunenthal had simply sent samples to doctors, asking them to try the drug and note the results. No-one reading the report of Grunenthal's 1955 symposium on thalidomide could guess that the company was employing some of the doctors who were reporting the best results. Grunenthal also played down reported side effects: dizziness, nausea and irreversible defects of the nervous system. To minimise medical scrutiny and increase profits, it promoted thalidomide both over the counter and on prescription.

When pressed, the company could not say whether the drug could cross the placenta (and hence perhaps affect the unborn child). Why not? Because Grunenthal had simply not tested the drug on pregnant animals. Sales boomed, but by 1959, there were also growing complaints from German doctors. Grunenthal hired a private detective to watch patients who complained, as well as officials and doctors who spoke against thalidomide.

By 1958, under licence from Grunenthal, Distillers were selling thalidomide in Britain — the blind leading the blind. They recommended its use in medicine as well as in obstetrics, where they said it was quite safe for pregnant women and nursing mothers.

In late 1960, the *British Medical Journal* (BMJ) printed a letter describing nerve damage from thalidomide. The next year, a BMJ advertisement assured doctors of its safety. But the same issue carried three more reports of nerve damage!

In the USA, after two years of investigation, one drug house decided not to market thalidomide. But another drug house, Richardson-Merrell, went ahead. They also assured doctors of thalidomide's safety, even for pregnant women, though they had not studied pregnant animals either.

The American "clinical trials" were just as shonky. Salesmen found over 1200 doctors willing to be "investigators", but told them "they need not report results if they don't want to". These doctors distributed over two million thalidomide tablets to 20 000 patients. There was no suggestion of informed consent, let alone any control group. When thalidomide was finally withdrawn, many of these investigating doctors had no records to allow anyone to trace the women! But the vigilance of Dr Frances Kelsey of the Food and Drug Administration (FDA) kept thalidomide off the general market in the USA.

Finally, in November 1961, a German professor alerted his colleagues.

The public outcry in Germany forced Grunenthal, after more than four years, to withdraw the dreadful drug. Still they denied any problems. But the English medical journal the *Lancet* now printed Dr McBride's famous letter from Sydney. This reported multiple severe abnormalities in almost 20% of babies whose mothers had taken thalidomide. Finally, the companies did the animal work they should have done in the first place years before. They found that thalidomide did cross the placenta in mice. Thirteen of 18 newborn rabbits showed the characteristic deformities of thalidomide babies. But in Britain, Distillers had still not given up; they kept trying to get thalidomide back on the market. In Japan, people could buy thalidomide for another year. A Swedish company continued to sell it in Argentina for three months after withdrawing it at home!

The most powerful chapter in the *Sunday Times* book *Suffer the Children* deals with families affected by thalidomide. One mother said: "When they gave [my baby] to me, his face was split, hanging apart like on a butcher's slab. The doctor was crying and said

my baby wouldn't live. But he did (blind and without arms) ...
I didn't cry outwardly, but inside I screamed."

At first, many families did not know the cause; many doctors
did not tell them. Some parents blamed themselves or each other;
there were suicides and broken marriages. One Belgian couple,
with help from their GP, poisoned their daughter who had been
born without legs. These parents were acquitted of murder.

Some nurses and doctors shunned the children. But there was
pressure on them to do something. So doctors operated up to 40
times on some children. Many of the operations helped, but many
did not. Some blind, deaf, dumb, paralysed and retarded children
are still alive. Some parents are still changing nappies on children
who are now in their 30s.

Just as sordid as the marketing of thalidomide was the
treatment of families claiming compensation. Only in Germany did
the government act on their behalf. In Britain, courtroom battles
pitted shattered, isolated people with poor legal advice against the
top talent hired by Distillers — a real case of David and Goliath.
Publicity might have helped the claimants, but for year after year,
fear of contempt of court restrained the media.

Dr John Newlinds was Medical Superintendent of Sydney's
Crown Street Hospital in 1960. He believes the book *Suffer the
Children* is essentially accurate. But was this tragedy preventible?
Dr Newlinds also pointed out that in 1960 doctors had not known
of any drugs causing major defects in the newborn. Moreoever, even
before thalidomide, phocomelia (absence of the upper arm and/or
upper leg) did occur (though rarely). He blames the companies for
not looking for such effects but is unsure whether animal testing
would have shown the danger to humans. The Australian Drug
Evaluation Committee was set up after the thalidomide tragedy.

There were at least 30 live thalidomide-affected children in
Australia. Around the world, there are perhaps 10 000. But how
many more unborn babies had defects which killed them?

(GB)

A PATIENT'S REVENGE

Each December, doctors look forward to a few Christmas cards and perhaps even a present or two from grateful patients. But patients aren't always grateful; some patients can turn on their doctors. In Brisbane, one man killed two orthopaedic surgeons who he felt had failed him.

Thirty-nine-year-old Karl Kast was a naturalised Australian. At 2.50 p.m. on 1 December 1955, he walked up the stairs of Wickham House in Brisbane's quiet tree-lined medical area and shot Dr Michael Gallagher three times. At least one shot entered the chest. It was only prolonged surgery that saved the doctor's life. Then Kast ran downstairs, lighting a gelignite bomb, with which he dashed out into the street. A passing horse-trainer tried to put out the fuse and lost three fingers in the process.

In Ballow Chambers, 100 metres away, Kast shot and killed Dr Arthur Meehan and Dr Andrew Murray. He also tried to grab Dr John Lahz, who suffered severe shock. The killer then locked himself into the consulting rooms. First came the sound of a shot; then the explosion of another gelignite bomb — two murders, a severe wounding and a suicide, all in about ten minutes.

The president of the British Medical Association's Queensland Branch called the deaths of Dr Murray and Dr Meehan "unparalleled in peacetime".

The next day Brisbane Criminal Investigation Branch got Kast's letter, presumably mailed just before the murders. He complained of the injustice of the specialists who had refused to give him worker's compensation for his back injury. Kast had intended to send all four doctors "into oblivion". Enclosed was a week-old clipping from Canada about a patient there who had shot his orthopaedic surgeon.

There were more explosives in Kast's boarding house room as well as another letter: "Seeing there is no justice and life is devoid of hope, I must look for justice elsewhere ..." Police could find no relatives. His personal effects went to his landlady, whose son described Kast as considerate and fond of children.

Born in Bavaria, Kast came to Australia in late 1938 and was interned when war broke out. A fellow-internee told police:

> He [Kast] was amazingly quiet and never had any friends
> .... he escaped once but was soon recaptured. [Once] he
> disappeared in the camp and we found him hiding under a
> hut.
>
> He was a brewer by trade, but out here he was made to do
> heavy work. He told me then that he had something wrong
> with his spine ... After the war we were sent to Alice Springs
> to work for the Civil Construction Corps ... again they tried to
> get him to do heavy work. Apparently specialists would not
> agree with him that he had a spinal injury and ... this kept
> building up inside him ... I think the authorities could have
> helped him, but probably, without realising, made it harder for
> him instead.

Only detectives, pressmen and gravediggers went to Kast's burial. Could anyone have reached out to Kast? Could three lives have been saved?

(GB)

Voyage of the damned: The Second Fleet

*The Second Fleet ... took many unsuitable convicts from
England's overcrowded and diseased gaols and hulks, put
them in unsuitable ships controlled by avaricious and
unsavoury men, and deposited those few who managed to
survive on an ill-prepared colony.*

COBCROFT, "MEDICAL ASPECTS OF THE SECOND FLEET",
AUSTRALIA'S QUEST FOR COLONIAL HEALTH

SECOND FLEET
Destination : AUST
Cargo : Unsuitable convicts

The Second Fleet ships used to transport convicts were not a nice way to learn about sea travel!

Three ships left Portsmouth on 19 January 1790 to bring 1000 convicts halfway around the world to Port Jackson. By then the colony was in desperate straits. The weekly ration was down to 2.5 lb of flour, 2 lb of weevilly rice and 2 lb of shrivelled salt pork.

On 3 June 1790 the colony joyously greeted not the Second Fleet, but *Lady Juliana*, the first convict ship to reach Australia since the First Fleet. She brought the first news for nearly three years: George III's madness, George Washington's inauguration, the French Revolution, and the reason why no supplies had arrived for so long: the *Guardian* had struck an iceberg.

But the crunch was that *Lady Juliana* not only brought little food, but dumped on the colony an "unnecessary and unprofitable" cargo of 222 women convicts. At least they were healthy, not like the prisoners on the Second Fleet (*Neptune, Surprize* and *Scarborough*) which soon followed. The Reverend Johnson boarded the *Surprize*: "Went down among the convicts ... a great number of them laying ... unable to turn or help themselves. The smell was so offensive I could scarcely bear it. ...Some of these unhappy people died after the ships came into the harbour ... "

At this time, naval officials in London were submitting to the King their proposed idyllic Great Seal of New South Wales. It showed "Convicts landed at Botany-Bay; their fetters taken off and received by Industry sitting on a Bale of Goods with her attributes, the distaff, bee-hive, pickaxe and spade, pointing to oxen ploughing, the rising habitations, and a church on a hill ... "

Primary sources on the Second Fleet disagree on the number of convicts and the death rates. But in round figures, of the 1000 prisoners transported, about one quarter died on the voyage, one half were landed sick, and one eighth died soon after. Only about one eighth were still alive and well some weeks after landing. The mortality on these three ships was the highest in the history of transportation.

Fatal Shore author Robert Hughes describes: "the starving prisoners ... chilled to the bone on soaked bedding, unexercised, crusted with salt, shit and vomit, festering with scurvy and boils".

What went so wrong? The Second Fleet was at sea for a shorter time than the First; the ships were well provisioned. Why then did so many die?

Many convicts were already sick when they were embarked from the hulks moored on the Thames. Scurvy, dysentery, vomiting and fevers were common, possibly louse-borne typhus as well: "One man ... had ten thousand lice swarming on his body." Drinking water, especially that from the Thames, often went bad long before the Cape of Good Hope.

The water that the *Surprize* took in often covered its wretched convicts to above their waists. The contractors used leg-irons from their slave-trading journeys. Captain William Hill, guard commander of the *Surprize,* described these as having a short bolt "...not more than three-quarters of a foot in length so that they could not extend either leg from the other more than an inch or two; thus fettered, it was impossible for them to move but at the risk of both legs being broken". Many died with the chains upon them. Fear of an uprising often led masters to keep their convicts below decks.

Profits lay in dead convicts. In theory, convicts received two-thirds of the Royal Navy ration. This covered bread, flour, beef, pork, peas, butter and rice. But in practice, convicts were often cheated of their food and half-starved. So they concealed the deaths of their fellow convicts so that they could have their rations; they took tobacco from the mouths of the dead; they stole food from the hogs.

Whereas the First Fleet had been a naval exercise, the Second was a private venture. The government chartered the ships from London merchants and paid them a flat rate of £17, 7s 6d for each convict embarked. Though contracts required merchandise to be landed safely, they said nothing about the condition of the convicts.

Captain Hill railed against the "villainy, oppression and shameful peculation" of the masters of the *Neptune* and *Scarborough*: "The more they can withhold from the unhappy wretches, the more provisions they have to dispose of . . .". On arrival, the masters made huge profits selling the withheld food.

News of the Second Fleet's fate caused public outrage in England. In the face of accusations by the crew of *Neptune,* the master (Donald Traill) and mate fled. The merchants also escaped prosecution.

Though the promised inquiry never happened, great improvements did follow. The payment system changed: contractors received £5 of their capitation fee only for those convicts who landed in good health. The navy started to appoint naval surgeons who were of a higher calibre than those available to the contractors.

Death rates on convict transports fell dramatically. On the *Neptune*, almost one in three convicts had died. But only one year later the *Salamander* lost fewer than one in 30!

(GB)

MAD KING LUDWIG II OF BAVARIA

In the introductions to Walt Disney's offerings for children which used to feature on television several years ago, those concerned with Fantasyland showed a castle with tall, slightly askew, turreted towers set in romanticised crags and forest. Despite the fairy glitter, it was in fact a fairly accurate representation of a real-life Bavarian castle, Neuschwanstein.

This imaginative place was built as recently as the 1880s by the capricious King Ludwig II of Bavaria, a monarch whose tenuous contact with reality was such that Disney had made an inspired choice to represent fantasy. While it is generally acknowledged that Ludwig was definitely odd, the question for medical historians was whether his actions amounted to true mental illness or whether he was just an eccentric who could afford to indulge his extravagant and romantic whims only to fall foul of psychiatric bungling on the grand scale.

It is an extraordinary story, so let's first recount his bizarre last few days and then look at his mental background.

Ludwig was born in Munich in 1845, the eldest son of King Maximilian II of Bavaria and Maria of Prussia. As a boy he received the strict Teutonic upbringing thought appropriate for the children of central European royalty. At the age of 13 he became enamoured by Richard Wagner's music and when he ascended to the throne in 1864 he sought out the composer. They formed a close relationship in which they spun fantastic plans for a new world of music until their union became a national scandal and Wagner had to leave the country. They remained friends, however, and Ludwig remains famous as being the composer's indulgent patron.

Following this estrangement, and after breaking a loveless engagement of convenience which he had undertaken with his cousin, Ludwig became concerned almost exclusively with artistic endeavours, developing a lavish passion for building castles of fantastic opulence in the Bavarian mountains. Once his whims were indulged his thought pattern changed and he became suspicious, arrogant and unpredictable towards his cabinet. He frequently managed to paralyse government by going off for long periods to one or other of his three castles. He was rarely seen by his subjects and was given to solitary brooding and long sleigh rides at night. His brother, Otto, was declared insane in 1875, and it was generally thought the King was going the same way.

Continually frustrated, early in 1886 the cabinet felt enough was enough and asked the Professor of Psychiatry at Munich, Bernhard von Gudden, to assess the King's mental health. As a personal examination was thought impossible, Gudden collected documentary evidence and the views of the staff to form an opinion. Without setting eyes on the subject he concluded that the ruler was indeed insane. This rather chancy path to diagnosis was picked up by the worldly Chancellor Bismarck who neatly dismissed the report as "tittle tattle and the rakings from the King's waste paper basket".

With the help of three other doctors but still no personal contact, Gudden reworded the document, and stated that none of the team had any doubt about the severity of the King's madness. All signed the report which concluded that Ludwig (1) was in an advanced stage of paranoia, (2) was incurable and would get worse, and (3) would forever be incapable of exercising government. Not much equivocation there, and one must marvel at their certainty, especially as the Bavarian Constitution stated that a monarch incapable for more than a year could be deposed. The report was submitted on 10 June 1886, and the immediate upshot was that the King's uncle was declared Regent.

The following day, a small medical posse headed by an assured Gudden left for Neuschwanstein with the dubious task of taking the patient, a complete stranger, into protective custody. The King had been forewarned of the situation and on their arrival had the group arrested. He then made a fatal error by ordering that they be flayed alive and then flogged to death. Such punitive excesses alerted the guard that their commander may indeed be ill, so the doctors were released.

The cabal left, regrouped and reappeared at 1 a.m. on 12 June. They were met by the King's valet who solemnly pronounced that Ludwig had been drinking heavily and wanted the keys to the castle's Fantasyland tower in order to commit suicide. That finally did it: the King was surrounded, confronted for the first time by Gudden, told of the findings, and immediately, and not unreasonably, responded by questioning the validity of the arms-length diagnosis. Gudden showed the documented proof, garnered from hearsay, and the patient was seized and taken away to Castle Berg.

The next day, 13 June, Gudden walked with the King through the castle park, presumably (and hopefully) to get some symptoms at first hand. They were followed by two attendants. All went well and another ambulatory chat was arranged for later that afternoon. At 6.30 p.m. the pair set off again, but due to some confusion in the orders from Gudden the two attendants returned to the castle rather than falling in at a discreet distance to the rear.

When the walkers had failed to return by 11 p.m. a search party was sent out. Both psychiatrist and monarch were found floating in the lake, dead.

The sequence of events is conjectural, but has been explained as going probably something like this.

While skirting the lake it seems likely that Ludwig had made a sudden dash for the water, which, incidentally, was quite shallow for 30 metres from the shore. Gudden must have held on to his coattails for one of his fingernails had been torn away. The King then seems to have slipped out of his jacket, turned on the luckless Gudden, and gripped him by the nape of the neck while sinking the fingers of his other hand into the psychiatrist's throat. He then held the head under water until the doctor drowned. All this has been deduced from the postmortem report.

Ludwig II, already in a suicidal frame of mind, walked a further 25 metres or so into the lake, where, still in shallow water, he managed to drown himself. His watch had stopped at 6.50 p.m. It seems on the evidence that the doctor was either trying to save the life of a patient hellbent on suicide or was stopping him from escaping across the two-kilometre-wide lake. Whatever the facts, he perished in the attempt. Gudden was 62 and Ludwig 40.

Despite the dubious psychiatric practice shown in this saga, Bernhard von Gudden was regarded as a man who, by the lights of the day, ran his unit in a humane way. It was he who had introduced to Germany a "no restraint" policy which allowed disturbed patients as much freedom as possible. It may have been this liberality which in the end was his undoing, although on the day of his death he had commented to other members of his party how dangerous the likes of Ludwig could be. The misunderstanding with the attendants and their premature departure made the difference between life and death.

After his presumed murder, there was little public sympathy for Gudden. Indeed, his grave was later desecrated and his widow was threatened with violence.

If that is the bizarre story of the royal death, what about the mental health of Ludwig and his family? Was the king really mad or just a romantic and self-indulgent eccentric?

His family, the Wittelsbachs, had ruled Bavaria for 700 years. Grandfather Ludwig I was a byword in both European whoredom, for maintaining the infamous Spanish dancer and adventuress, Lola Montez, as his mistress, and civic largess by dint of unrestrained spending on some splendid public buildings in Munich.

Otto, Ludwig's younger brother, was pronounced incurably insane at the age of 25. He is supposed to have been a florid schizophrenic, but in letters to his old governess Ludwig revealed that. "...Otto makes terrible faces, barks like a dog and sometimes says the most indecent things, and then again he is perfectly normal for a while". Gilles de la Tourette, the Parisian neurologist, was not to describe his famous syndrome until nine years later in 1884, but these royal observations sound almost better than his classic description of a malady to which he gave its eponymous title. In any event, Otto was kept confined for 30 years until his death in 1916.

Their cousin, Rudolph, Crown Prince of Austria, was to shoot himself in 1889. He seems to have entered into a suicide pact with a lover, for he was found in bed with a naked 17-year-old girl also dead from gunshot wounds. Ludwig's relatives from Hanover, then on the throne of England in the embodiment of Queen Victoria, the last of the Hanover line, are postulated to have been tainted by porphyria, a malady which can become evident as severe mental disturbance; it had already famously manifested itself in King George III. And finally a paternal aunt was once kept in an asylum while suffering under the unshakeable and remarkable delusion that she had swallowed a glass grand piano. So all in all Ludwig's psychical background was not propitious.

Ludwig grew up a shy and highly strung boy whose homosexual propensities soon became evident. He had romantic attachments to several aides, troopers and stable lads, many of whom later sold his revealing letters. Nonetheless, at the age of 22 he became engaged to his cousin, Sophia, but the arrangements were called off when his relationship with an equerry, Richard Hornig, burgeoned. It was to last for years. Following his broken engagement, Ludwig became withdrawn and remote, suffering great feelings of guilt about these homosexual urges.

He retired to Castle Berg where he developed the wearisome habit of turning night into day. His best-known quirk was to imagine he was riding to a distant town, estimate the number of circuits of the riding school it would take to make the distance, then proceed to do it through the night, taking appropriate breathers at phantom inns. The Crown Princess of Prussia, the eminently sensible daughter of Queen Victoria, reported that he dined with a horse which was kitted out with a gold crown. Not unnaturally, people began to talk.

From 1869 until his death in 1886, Ludwig kept a diary, two volumes of which were eventually published. As early as 1870 ruminations about suicide were committed to the book, such as "The waters of the Alpsee beckon to me". But many entries centre on mastering his guilt about his homosexuality, and on unbridled feelings of majesty and preoccupation with Louis XIV, the so-called French Sun King who died in 1715, 150 years previously. He was fascinated by the mystical significance of numbers and in one of his written pieces made the nonsensical conclusion that "10 + 11 = 21, the number of terrible memory, but quite exceptionally 10 + 11 make 12". There were many other tortuous numerical ruminations which are surely highly significant in the thought disorders of schizophrenia.

Through the 1870s he became even more preoccupied with his castle building, mostly on inaccessible crags. If his ministers visited he sat behind a screen, and although unseen, it was admitted that some of his observations and questions were shrewd enough. For the last three years of his life, and after the death of his romantic idol, Wagner, he became a complete recluse. In the composer the King had found a musical genius with an imagination extravagant enough to match his own.

It was thus in 1886 that Bernhard von Gudden was eventually asked to report on the King's mental state, and, as we have seen, this was accomplished by merely the logging of depositions. Indeed, some Wagnerian scholars, music critics and other apologists have since called the exercise "a political conspiracy done by medical hirelings", and the conclusion of "paranoia" used to cover what was only "a multitude of minor eccentricities".

Richard Hornig, his lover, was very reluctant to testify, but others had no such reservations. The sergeant of the Royal Stud told how when out walking Ludwig said he could hear voices, could never identify the words, but would apparently answer them. His valet testified the King imagined things and could not be dissuaded in this belief. For instance, after he ordered a knife to be removed from the dinner table, when it was obvious to all that the table was bare, he ordered an hour-long search for it to be carried out. Staff told of outbursts of rage, or hours of mute contemplation, and thirty-two servants reported maltreatment in some way. There could have been some axe–grinding here, of course, so corroboration would be needed. Although he always ate alone, all dishes had to be prepared for four people as the King imagined he was entertaining guests, a fiction he perpetuated by carrying on an imaginary four-way conversation.

From various accounts it seems the first signs of illness were as early as 1865 when he was 20. Definite signs of withdrawal had appeared by 1867 when the engagement was broken off, but the florid mental symptoms did not manifest themselves until about 1883.

Could he have been another right royal example of porphyria, the scourge of the Hanoverians? What about the central nervous system manifestation of tertiary syphilis, grandly styled by doctors as general paralysis of the insane or GPI; perhaps as a distant link with the exotic Lola? Did he have a toxic psychosis due to alcohol, of which it is known he was fond? Or were his extravagant quirks

First, thanks for coming to dinner. Can you please pass the salt?

King Ludwig always ate alone, but insisted on having dishes prepared for four people. His imaginary friends were some of the most well-fed of all imaginary friends!

after all merely whimsical self-indulgences carried out by an autocrat who could afford it?

I think none of these. Despite the medical bungling of the certification — not to have seen the patient would rightly be impossible now — I am sure Gudden got it right. The age of onset, the family history, the delusions and hallucinations observed by people whose loyalty was never in doubt, and, crucially, the self-recorded thought disorder, all point to florid schizophrenia.

And yet in the whole sorry fandango, almost as tragic a figure as Ludwig was the doctor, Bernhard von Gudden. He has been portrayed as the villain of the piece, but almost certainly he died trying to save his patient's life. It was just one more aberrant twist in an already bizarre case.

(JL)

SAUERBRUCH: THE RISE AND FALL OF A GREAT SURGEON

In *The Doctor's Dilemma*, George Bernard Shaw called medicine and other professions "conspiracies against the laity".

In *The Dismissal: The Last Days of Ferdinand Sauerbruch, Surgeon*, Jurgen Thorwald writes, "Doctors feel obliged to cover up professional errors on the part of fellow-physicians ... Professional discretion shields the bungler".

On 17 April 1951, Dr Ferdinand Sauerbruch (1875–1951) operated on 41-year-old Berlin woman Irmgard Fiebig. Deep inside her throat, she had a secondary cancer the size of a hen's egg.

He had already operated in 1947 for her breast cancer. But towards the end of 1950, she had noticed a lump in her neck. Then it became painful. After ignoring it for some time, she returned to the clinic where she had first seen Sauerbruch, but the staff there turned her away. Finally, she talked her way into his house to see Sauerbruch there. Hence her second operation, one of the most bizarre catastrophes in the history of surgery.

In his own home, with only local anaesthetic, Sauerbruch started with a cut of about 20 centimetres in her neck! She endured agonising pain, not only during the operation itself, but for the rest of her life. Finally, six months later, Irmgard Fiebig won release in death. Why did she suffer thus? Why did a surgeon of international renown attempt major surgery in this barbaric way?

Way back in 1904, as a young doctor, Sauerbruch had made a huge breakthrough by finding a means by which surgeons could open the chest. Thus he "put an end to a centuries-old medical doctrine that chest diseases could not be treated surgically". His methods allowed surgeons to operate on the lungs and heart and led to great advances in the treatment of tuberculosis and cancers. He lectured in America and worked in the leading hospitals of various European capitals. At the Charité Hospital in Berlin, his surgical unit brought great renown to German medicine.

But Sauerbruch was a heavy drinker and a philanderer. According to Robert Youngson and Ian Schott, his behaviour was "a bewildering hotchpotch of kindness and harshness, wisdom and thoughtlessness, great compassion and explosive temper, irresistible charm and cold hauteur ... unaffected modesty and extraordinary conceit". Such erratic behaviour "made it difficult to contend with Sauerbruch when he lost his faculties and began to kill patients".

After the Second World War, Berlin was divided. Sauerbruch, though in his 70s, still headed surgery at the Charité hospital, which belonged to the Soviet-controlled sector. To Soviet doctors, he was a hero.

Dr Friedrich Hall, who was in charge of medicine in East Berlin, first met Sauerbruch in 1948. It struck Hall that Sauerbruch, despite his reputation for mental vigour, repeated himself, forgot names and talked of dead people as though they still lived. Then Hall came across tapes of Sauerbruch's lectures given before the war. The contrast convinced him that the old man now had dementia. That made him "as dangerous with the knife as a drunken man". Indeed, by now Sauerbruch had already caused at least one needless death. In 1946, he operated for a hernia on a private patient of 39 years. That night, the man bled to death.

Hoping that the old man could not keep working much longer, Hall just waited. But soon after that Hall heard that a respected surgeon, Karl Stompfe, had suddenly resigned after Sauerbruch, his chief, had hit him. Then, in Hall's presence, another assistant reported to Sauerbruch that he had found an inoperable brain tumour. Sauerbruch rushed into the operating theatre. A few minutes later, he held out his bare unwashed hand to show Hall the tumour: "Look at that ... I went right in and pulled it out. The finger is still the best instrument". So another patient died.

Hall now asked the professor of pathology about unexpected deaths at the Charité. Though the pathologist was an old friend of Sauerbruch, he admitted that Sauerbruch was dangerous.

Still the authorities insisted on retaining him to lend status to the now communist-controlled Charité! The party line was: "In the coming struggle of the proletariat, in the clash between socialism and capitalism, millions will lose their lives ... it is a trivial matter whether Sauerbruch kills a few dozen people on his operating table. We need the name of Sauerbruch."

Late in 1949, Sauerbruch operated on a boy with stomach cancer and closed the bowel and the stomach without rejoining the two — another death.

Finally, two of his superiors invited Sauerbruch to supper. Over three agonising hours, they appealed to his pride and finally persuaded him to resign rather than be dismissed. On 3 December 1949, Sauerbruch did officially resign from the Charité. But then he still kept coming each day to work, until they had to lock him out of his office.

Sauerbruch was flattered to receive speaking invitations from colleagues who felt he had been unjustly treated. The medicos of Hanover turned out to hear him speak on the philosophy of medicine. But the demented old man lost his thread; soon he could not string a sensible sentence together. He himself was happy with his speech, but his shocked colleagues begged the reporters to hush up the fiasco. Next Sauerbruch started to work at a private clinic, until his employer realised his error. Each day he would tell Sauerbruch that there were no patients for him, or that the theatre was booked out. This farce continued for months. Then, in 1950 came the most tragic phase: the old man trying to operate in his own home, where some devoted old patients tracked him down. Time and again, the authorities avoided their responsibilities by asking his wife, Margot, herself a doctor, to stop her husband. Margot disconnected the doorbell, since she had to work and could not afford to stay at home to watch over the demented old man. That was how, in 1951, Sauerbruch operated on poor Irmgard Fiebig. Finally, the authorities banned him from seeing any patients at all.

The tragedy did not end until Sauerbruch himself died a pauper on 2 July 1951. His ghosted memoirs, *That Was My Life*, finally published after his death, were remarkably successful, selling almost a million copies.

(GB)

MEDICAL BOARD CRUCIFIED GP

Doctors who received an orthodox medical education did not look kindly on unqualified practitioners, especially those who cured the medical profession's failures.

MALTBY & LEE, "THE MEDICAL ESTABLISHMENT AND ASSOCIATION WITH UNQUALIFIED PRACTITIONERS: THE SAD CASE OF DOCTOR AXHAM", *JOURNAL OF MEDICAL BIOGRAPHY*

About a hundred years ago, Herbert Barker was at London's Paddington station bound for a holiday on Jersey Island. Without ceremony, a total stranger limped up to him: "I've come all the way from Sydney. I've been chasing you from pillar to post." Barker pointed out that his train was about to leave. "You could at least examine me on the train," the Australian persisted.

By the time they were out of London, Barker had offered to manipulate the visitor's displaced joint (we don't know which joint) as soon as he returned from his holiday. The man said, "Do it now, I don't mind pain." Swaying with the speeding train, Barker braced himself and manipulated the joint. Later he wrote: "This man ... had to travel half the globe's circumference ... all just to get put right an injury that any man properly instructed in manipulative surgery should have been able to handle."

Born in 1869, Herbert Barker first served as apprentice to his cousin, a well-known but medically unqualified bonesetter. Later, he inherited the practice and ended up in Park Lane, where his patients included H. G. Wells, the Duke of Kent and George Bernard Shaw. Barker's manipulation often relieved dodgy necks and knee joints, back strain or painful feet.

Dr Frederick Axham, a Soho GP, found that Barker could help some of his (Axham's) patients when he (Axham) could not. To spare patients the pain of manipulation, Axham gave anaesthetics for Barker from 1906 to 1911. No problem.

But in 1911, a patient had a complication after Barker manipulated his knee. When Axham testified for Barker, the Medical Defence Union complained to the General Medical Council (GMC) about Axham giving anaesthetics for Barker. The GMC charged Axham: "That you have knowingly and willingly ... assisted ... an unregistered person in a department of surgery, in carrying on such practice ... and [are] guilty of infamous professional conduct."

After Axham refused to cut his ties with Barker, the GMC ignored his unblemished 49 years of practice and deregistered him. No appeal was possible, but his letter appeared in *The Times*:

> *My professional status is stripped from me for an association which was and is desirable ... It is undesirable that Mr Barker should operate without an anaesthetic ... the alternative to a qualified man is an unqualified man. Convinced ... that Mr Barker's methods are both sane and sound ... I had no*

hesitation in associating myself with him in a work which has resulted in unspeakable relief to thousands of patients whom surgeons by orthodox methods absolutely failed to relieve.

Though now unregistered, Axham continued to give anaesthetics for Barker but could no longer practise in any other way. Over the years, Axham's friends, as well as Barker and his friends, all tried to get Axham reregistered. So did the press: "The ... profession has a habit of crucifying its most gifted members on the cross of conservative prejudice." All in vain.

After Axham retired altogether in 1921, many younger doctors offered to give Barker's anaesthetics. The chosen one was Dr Frank Collie, from an eminent medical family. He wondered if he would also be struck off. But the GMC's silence was deafening. As Barker wrote, "the law had a Nelson eye".

Sir William Arbuthnot Lane was among the prominent surgeons whom Barker won over. Lane told his students at Guy's: "The bonesetter ... has profited by the inexperience of the profession and by the tendency ... of adhering blindly to creeds whose only claim to consideration is their antiquity."

As Barker's star rose, so did Axham's fall; he lived in genteel poverty and professional disgrace. In 1922, Barker won a knighthood for his services to manipulative surgery. Four years later, Axham died at the age of 86. His last words were: "I forgive as I hope to be forgiven." Herbert Barker paid tribute: "All who knew the man and know the facts hold my old friend and colleague in the highest esteem."

Professor Roger Maltby remembers hearing that in the 1920s his own mother, a physical education student, had a locked knee. After a surgeon put it in a plaster cast for six weeks, her parents took her to Barker, who not only unlocked the knee, but showed her how to do it herself. He charged her only 30 guineas (the same fee he charged footballers), whereas most patients paid 100 guineas.

In 1936, the British Orthopaedic Association invited Barker to show his skills at St Thomas's Hospital. The president-elect of the

Association of Anaesthetists won thanks for anaesthetising! But Barker, still thinking of his old friend Axham, challenged the doctors: "Gentlemen, are you not now guilty of unprofessional conduct?"

Dr Rowley Bristow was among the 100 surgeons there. He wrote in the *British Medical Journal*: "Had [Barker's] offers to demonstrate ... been accepted twenty-five years ago, the general utilisation of this branch of therapeutics (manipulation) would not have been so long delayed."

Drs Maltby and Lee sum up: "The GMC could have reviewed its earlier disciplinary action against Axham before he died; in view of its omission to take similar action against his successors, its failure to do so appears both hypocritical and unjust."

A comment from Dr Peter Arnold, veteran of Australian medical politics: "A latter-day Dr Axham who gave anaesthetics for manipulative (as opposed to operative) surgery would have no problem with today's NSW Medical Board."

(GB)

THE FREEZING OF FRANKLIN

In a 12-day voyage during September 1969 the 151 000-ton oil-tanker *Manhattan* traversed the 1000-kilometre Northwest Passage off northern Canada. The task was accomplished with contemptuous ease and the *Manhattan* became the first ship to make its way through the five-metre thick ice of the passage and be back at the fleshpots of New York within the month.

This was not, however, the first vessel to make the transit. That single honour belongs to the ship of the Norwegian explorer Roald Amundsen who, setting off in 1903, took three years over it, navigating only in summer to eventually finish in 1906. He did it in a 47-ton herring boat and dallied, it is said, to escape his creditors.

The first single-season voyage was in 1944 by Sergeant Larsen of the Royal Canadian Mounted Police.

The original object of this seemingly foolhardy mission was to find a quick way to the Orient from Europe, a desirable and time-sparing objective of the pre-flying days and which had first been mooted to the well-regarded navigator John Cabot, by King Henry VII of England. That was in 1497, and between then and 1906 there were many disastrous, albeit gallant, attempts.

But in all that time there were none more disastrous or more gallant than the expedition mounted in 1845 by a former Governor of Van Diemen's Land, Sir John Franklin. Van Diemen's Land is, of course, now called Tasmania, and Franklin is probably better known to the locals as having lent his name to their most famous river. But to the world at large, his exploratory zeal in the Canadian Arctic has become a byword in courage. And it is not without considerable interest in the annals of medical history.

The facts surrounding the start of this voyage of discovery are well documented. Of how on 19 May 1845 the naval ships HMS *Erebus* and HMS *Terror* sailed out of the Thames carrying 134 officers and men on the best-equipped expedition ever to leave for polar regions. The ships were the most technologically sophisticated vessels afloat and were built to go where no ships had gone before. There were provisions for five years, which at a pinch could have been stretched to seven, and they had retractable propellers, desalinaters, an internal heating system and some of the world's first cameras to record the epic event. They even had on board a wardrobe of costumes for amateur dramatics which had been planned to amuse the men during the long nights.

And of the expedition's leaders, the President of the Royal Geographical Society, catching the mood of the day, proudly stated, "The name of Franklin alone is, indeed, a national guarantee."

Freshly painted in yellow and black, the ships departed from Greenwich and were given a spirited farewell by 10 000 people, brass bands and the good wishes of a clutch of civic dignitaries. Five men were dropped off in Greenland and the ships with 129 souls on board were seen to enter Baffin Bay at the end of July. But then ... silence. Not one of them was ever seen alive again.

By 1847 the Admiralty began to wonder aloud about their fate. But not until 1850 did any urgency creep into plans for sending out a search party. During the subsequent hunt traces of huts were found, and also, most excitingly, three graves. A chiselled headstone put the earliest death at 1 January 1846. It was speculated that scurvy was the cause, but the inferior quality of the tinned food was considered a factor.

It was not until May 1859, fourteen years after their dazzling departure, that, together with a lifeboat, two skeletons and some abandoned equipment, the first and only written record was found. With agonising brevity it stated that Sir John Franklin had died on 11 June 1847 and the ships were abandoned on 22 April 1848, by which date nine officers and 15 men had perished. What it did not say was how they had perished and where the rest had gone. The observation of the search party leader, Captain M'Clintock, could not have been bettered when he wrote at the time, "So sad a tale was never found in fewer words."

Parts of skulls and other fragments of cloth and rope were found over the years, but interest languished until in 1981, 133 years on, Owen Beattie, an anthropologist from the University of Alberta, thought that King William Island, site of the known landing, might still hold some secrets of the Franklin disaster. The distillation of his findings over the following four or five years, which have been written up mainly by his co-explorer, John Geiger, have produced a unique scientific record of quite remarkable fascination.

The visits Beattie paid to the area and the sometimes grisly finds he chanced upon are not only documented with punctilious detail and literary grace, but are enlivened by a set of the most brilliant photographs; no petrified grin is avoided, no femur is left unturned. The account, however, is not for the squeamish or fainthearted. Several exploratory trips were made to the area of the disaster, and it eventually became evident that to get some sense into the mystery of the deaths, three main questions needed to be answered.

First, why was it that when a dismembered skeleton was found the skull fragments were near the campsite and the limb bones farther afield? Second, was it not strange that quite a few of the crew died fairly early on in the expedition, and, more than that, there was a disproportionate number of officers to men, including the leader? And lastly, and crucially, why was it that lead levels in the bones of crew members were ten times the levels of those found in Inuit from the same area?

Beattie originally thought these heavy metal readings were a chance finding, as, along with everyone else, he considered that the sailors had died of a combination of starvation and scurvy due to vitamin C deficiency. But to ensure completeness of any postmortem examinations, he requested that the laboratory do a blanket scan of all elements. This was unusual and time-consuming, but the unexpected results demanded a rethink of the situation.

The conclusion regarding the scatter of bones was the unsavoury one that during the last days cannibalism had taken place. While strength lasted, and perhaps to avoid the haunting gaze of your fellow sufferers, the more meaty limbs could be picked up and carried off to be eaten away from the tents. To crack open and eat a face and brains, it was reasoned, you would have to have been pretty hungry, insensate or getting to the end of your tether, probably all three. Therefore, you were more likely to be lying in the campsite, too weak to move far.

It was concluded that the answers to the second and third questions were related and opened up completely fresh theories from those put forward over the years.

The reasoning went that the contractor for the supply of the huge order of 8000 of the then relatively new-fangled tinned foods (remember this was in the mid-nineteenth century) had difficulty meeting the sailing deadline. He was Stephen Goldner, a victualler from London. To expedite delivery he employed unskilled labour, paid scant attention to hygiene and packed poor meat together with animal hair and bones. The most generous comment is that quality control suffered, but, furthermore and tragically, records were to show that a significant proportion of the food went bad due to indifferent sealing.

But there was even more to it than that.

At the time the tins were made from a wrought iron sheet which was bent into a cylinder and then soldered with lead both inside and out. Beattie collected a number of the cans which had been dumped on the shore and were still scattered about, and came to the conclusion that lead contamination from the soft solder, known to contain 90% lead, would have been quite considerable. The average amount of lead in the ordinary urban population is 5–14 parts per million, and symptoms of poisoning begin to develop after the absorption of about 40–50 parts per million. In recovered hair from exhumed bodies and other tissue, the figures ranged from 138–657 parts per million. It was concluded that lead poisoning and not just starvation was the prime cause of death.

Why more officers? Well, they would have had a preferred diet of more tinned food and less blubber, pemmican and the like garnered from the surrounding terrain. They also ate off lead-enriched pewter plates. For once class distinction worked to a disadvantage.

Three of the sailors were exhumed, a task not without its difficulties in view of the fact that the ground was frozen solid with permafrost. In the unlikely event that you are ever called upon to do the same, the trick is to pour boiling water on the soil and be prepared for a prolonged operation of tedious, unrewarding digging.

The drama was heightened by the fact that a distant relative of one of the men answered an advertisement the organisers had placed in a British newspaper to assist the expedition, sailed with the scientists and in the end came face to face with an ancestor whose likeness was in almost mint condition, preserved by the ice. It must have been an emotional moment for him.

As well as the excess lead, tuberculosis was a feature in those exhumed. This was a common disease back then and would not be the result of any privations suffered.

So I think it can be concluded that the long-held belief that vitamin deficiency was the prime cause of death is not true. It was more likely to have been lead poisoning and starvation, complicated by the then common and debilitating disease of tuberculosis. However, whatever the cause, slow death in such a solitary place, in bitter weather and knowing there was no way out, must have been particularly pitiable.

Franklin's nephew, Alfred Lord Tennyson, wrote the explorer's touching epitaph for his Westminster Abbey plaque. It goes:

> Not here: the white North hath thy bones, and thou
> Heroic Sailor Soul
> Art passing on thy happier voyage now
> Towards no earthly pole

But for these two remarkable and doggedly persistent Canadian researchers another quotation from the same orotund author could well apply. It is from his poem *Ulysses* and appears on the memorial of the gallant South Polar explorer Robert Falcon Scott. It goes, you will recall:

> To strive, to seek, to find and not to yield.

(JL)

AUSTRALIAN DOCTORS' BATTLE WITH LEAD POISONING

Medical researchers often criticise the delay between a key discovery and its widespread application: the gap between knowledge and action. One striking example is the prevention of lead poisoning. In this field, Australian research ranks with the best in the world, but we have been slow to apply this knowledge.

Hippocrates, Galen and Celsus were among early authors who described lead poisoning in adults. But only much later did doctors studying the families of leadworkers recognise the very different picture in children.

In the 1890s, there was an epidemic, mainly in Queensland and northern New South Wales, of children with anaemia, wasting, poor vision and stomach pain. Brisbane doctors studied 79 children, of whom many died. They showed that these children had lead poisoning. But how were the children exposed to lead? The intense search covered roofs, water tanks, pipes, guttering, cooking pots, home canning and even the silver paper wrapping of sweets. (This was before the days of leaded petrol.)

Then one hot Sunday an idea came to Dr Lockhart Gibson while he was relaxing on his verandah. He saw that the paint on his railings was powdery. After he wiped this powder onto a clean rag, the government analyst confirmed its high lead content! On the woodwork in the homes of the sick children, he found the same white dust. Thus doctors confirmed that peeling and flaking household paint was the source of the lead. This tied in with the higher frequency of lead poisoning during the Queensland summer, when paint became dry and powdery.

But why were only some children poisoned when whole families lived in painted houses? One mother told Dr Gibson that of her four children, the only ones to get lead poisoning were the two "nailbiters and fingersuckers".

The doctors warned the government of the dangers of white lead paint and pressed for legislation: "... recurrent attacks of lead poisoning could be prevented by getting them (parents) to replace lead paint by zinc paint on verandah railings and outside surfaces within the reach of children".

In 1929 doctors noted that chronic kidney disease was very common among young adults in Queensland and linked this condition with lead exposure during childhood. A special report in the *Medical Journal of Australia* pressed for "Legislation against lead paint ... on any articles available to children — furniture, toys, pencils and so forth, as well as on external surfaces such as verandahs."

But still nothing changed. Further reports hammered home the same message. Finally in 1955, the government banned lead from all paint made, sold or used in Queensland. But that was not the end of the problem, either in Queensland or elsewhere. Even now many people live in houses containing old, lead-based paint. It will still be a long time before we see the end of poisoning by lead in paints in Australia.

(GB)

The murky history of tobacco

Have you heard of the man who got so worried when he read all the articles linking smoking to cancer that he gave up reading?

Most authorities agree that the use of tobacco started in South America. For many centuries before white exploration, indigenous Americans grew and smoked tobacco. They used it in ceremonies, for its intoxicating effects and to ease aches and pains, snakebite, chills, fatigue, hunger and thirst. They smoked it in cigars, pipes and cigarettes (wrapped in cornhusks); they made tobacco syrup to swallow or apply to the gums. They chewed it or snuffed it. The Maya of Yucatan (now in Mexico) regarded tobacco smoke as divine incense to bring rain. Perhaps the oldest representation of a smoker is a stone carving from a Mayan temple showing a priest puffing on a ceremonial pipe.

Christopher Columbus was the first known white man to meet the weed on the island of Cuba, or perhaps San Salvador. One of his crew brought the habit home. But back in Portugal, his friends, seeing smoke pouring out of his mouth, were sure he was possessed by the devil. He languished in jail for several years.

By the early sixteenth century, Spaniards in the West Indies and Portuguese in Brazil were smoking more and more. Most could not stop.

In 1559, Jean Nicot, French ambassador to Lisbon, used tobacco leaves to cure his cook's cut finger. Then he sent seeds as medicine

Christopher Columbus met up with the weed via a crew member venturing through Cuban islands. Back in Portugal, folk seeing smoke wafting from his mouth were sure that he was possessed by the devil!

to the French Queen Catherine di Medici, who grew medicinal herbs. *Nicotiana tabacum* is named after him. It is related to potatoes, belladonna and henbane.

In the 1580s, Walter Raleigh set up a colony in Virginia; Francis Drake brought back the tobacco habit. It is reported that Sir Walter Raleigh's servant, on first seeing smoke drifting from his master's mouth, poured a bucket of water over him!

Many of the large tobacco plantations around the world got their start from the Portuguese. By 1600, all the maritime nations of Europe were using tobacco. But in England, King James I, who succeeded Elizabeth I, challenged its popularity. He had his *Counterblast to Tobacco* published anonymously:

*A custome lothesome to the eye, hateful to the nose, harmfull
to the braine, dangerous to the lungs, and in the black stinking
fume thereof neerest resembling the horrible stigian smoke of
the pit that is bottomless.*

James decreed that people who were caught smoking would
have the pipe rammed down their throat — an early example of
health education. Next James organised at Oxford the first public
debate on tobacco, at which he showed black brains and innards
said to come from deceased smokers. He increased the tobacco tax
40-fold, but this just made smuggling boom.

In the 1600s, Turkey, Russia and China imposed (and enforced)
the death penalty for smoking. Yet the Chinese became the heaviest
smokers of tobacco in Asia; later this led them to smoking opium.

In his 1653 work *English Physician* the herbalist Nicholas
Culpeper listed the many benefits of homegrown tobacco. People
used it in cigars or pipes or as snuff; some used tobacco as an enema
to revive the drowned. With special bellows, fans could even have
smoke enemas. One enthusiast pushed it as an infallible cure for 36
different ailments. Some people smoked to cure asthma, while
others took infusions of tobacco. Some used tobacco as a stimulant,
while others liked it as a sedative. Others swore by finely powdered
tobacco as a disinfectant. When plague broke out in London,
doctors said smokers were less likely than others to catch it. During
epidemics, scholars at Eton were thrashed if they *did not* smoke!
Doctors used tobacco as well as mercury in sweating baths for
syphilis. London had 7000 tobacco shops. While the poor mostly
used pipes, the upper crust also snuffed or sniffed tobacco dust.
Though various popes forbade its use, not even priests could stop.

It was over two centuries ago that some doctors warned patients
of the link between tobacco and cancer. In 1761, the English
physician John Hill described perhaps the first case of tobacco-
induced cancer: his patient had cancer of the nose. In 1851, the
eminent surgeon James Paget warned a man with a white patch
("smoker's patch") of the tongue where he rested the end of his

pipe: "he certainly would have cancer of the tongue if he kept smoking". Soon after, a French doctor reported 68 cases of cancer of the mouth; 66 of these patients smoked tobacco, while the other two used to chew it.

How did cigarettes come into existence? Beggars in Seville picked up discarded cigar butts, shredded them and rolled them in scraps of paper. Cigar makers, fearing the competition, spread rumours that cigarettes contained opium, and were made with tobacco from discarded cigar butts and paper made by Chinese lepers!

Warfare and smoking have often gone together. During the American War of Independence, George Washington begged: "If you can't send money, send tobacco." The French used to buy tobacco from the colonists, so General Cornwallis did his best to destroy the Virginia tobacco plantations. As long as chewing tobacco (usually a mixture of molasses and tobacco) was popular in the USA, all public buildings had to have spittoons.

The advent of cigarette-making machines and safety matches around the late 1800s set the stage for smoking to boom.

Soon after the First World War, doctors in Britain noted that lung cancer was becoming much more common. During the next half century, the annual death rate from lung cancer shot up. Scientists suspected an environmental cause: suspects included occupational exposure, urban pollution, tarred roads, exhaust fumes from internal combustion engines, and tobacco smoke.

Cigarette smoking among men boomed during the Second World War. In Britain after the war, Dr Richard Doll asked 600 patients with lung cancer about their smoking habits. Almost all were cigarette smokers, whereas there were fewer smokers among the control subjects (people without lung cancer). At this time, the male smoking rate in Britain was still about 85%; many people smoked to clear their lungs! Dr Doll was himself a smoker but now the evidence induced him to stop.

To nail down the connection, he needed to follow people in a prospective study. He sent questionnaires to the 60 000 doctors then practising in Britain. After five years he found "a marked and steady increase in the death rate from lung cancer as the amount smoked increases".

The good news: those who stopped smoking cigarettes reduced their risk. One eminent chest physician advised colleagues to follow his example. To keep his hands occupied after he stopped smoking, he took up knitting!

Sir Macfarlane Burnet, an icon of Australian medicine, also used to smoke. But by 1955, he gave up, saying, "The evidence [against it] is too strong." Moreover, he kept speaking up in public, urging Australians to do the same and the television stations to stop cigarette advertising. In 1971, he appeared in an advertisement for the Anti-Cancer Council of Victoria. Channel 7 showed the advertisement for a week, but then pulled the plug. The publicity following on from this censorship greatly advanced the goals of the anti-smoking lobby.

In 1962 the Royal College of Physicians in the United Kingdom stated clearly that smoking caused lung cancer. Two years later, the American Surgeon-General said the same thing. Was it this report that led activists in New York to carry scissors and cut cigarettes out of the mouths of passers-by smoking in the streets?

Long-term results of Doll's study on doctors showed that about half of all regular cigarette smokers would eventually die from a smoking-related cause. (Tobacco causes many diseases apart from lung cancer.)

In 1979 came a second report from the US Surgeon-General with further alarming findings: pregnant mothers who smoke may harm their babies; as smoking levels among women rise, so do disease and death rates ("Women who smoke like men die like men who smoke"); and more children were smoking.

The *Medical Journal of Australia* asked doctors to set an example by not smoking. Indeed, by 1982–83, only about 11% of Australian doctors were smoking cigarettes, compared with over 30% of the Australian population.

It is now hard to deny the devastating effects of cigarette

The rather annoying (for smokers) New York 1962
habit whereby non-smokers would carry scissors and
cut cigarettes out of the mouths of passers-by.

smoking on health. But as anti-smoking movements have hobbled
promotion in wealthier countries, tobacco companies have turned
to other markets.

While visiting Australia in 1998, Sir Richard Doll, then an
active 85-year-old, said, "Exposing the hazards of tobacco smoke is
the thing for which I would like to be remembered." He reserved
his harshest scorn not for tobacco companies, but for governments
that have failed to restrict tobacco promotion.

Even though the NSW government has just banned smoking in
indoor restaurants, cafes, shopping centres and casinos, and
governments in Australia have been among the global pioneers in
tobacco control, we are still a long way from achieving a smoke-free
generation in Australia.

The epidemic of death and disease caused by tobacco is still mounting all over the world. Tens of millions who smoke today are doomed to be killed before their time unless they give up the habit.

(GB)

Chapter 10

QUIRKS AND ODDITIES

IT'S NEVER TOO LATE TO CHANGE

Does life feel like one big drag? Do you want to do something quite different? One middle-aged doctor left his practice to sell hot dogs. His therapist had similar feelings of rebellion. This story struck a chord with me too. I was over 50 before I realised my teenage dream of writing.

Back to the discontented doctor and his family. When seven-year-old Charlie was referred to Dr Herbert Strean, a psychoanalyst in New York, his parents were not hopeful. Charlie was unruly and aggressive. He would flare up whenever he could not get his own way.

His mother and father saw themselves as model parents and felt that his bad behaviour reflected on them. The father, Saul, turned out to be a physician. Once when they were alone together, Dr Strean asked Saul how Charlie's behaviour affected him. Saul sprouted a long speech, saying that children should obey their parents: "When I was a boy, my parents' rules were like law. I was always good."

Over the next few months, Saul talked about his own childhood. Finally, he told Dr Strean a secret: "Sometimes I feel like giggling when Charlie is naughty." With a lot of prodding from the therapist, he went on: "I would have loved to be a rebel too, but I always did the right thing. I really never wanted to be doctor at all, but that was my parents' dream. I've never enjoyed my work."

"So what did you want to do?"

"I've always wanted to run a hot-dog stand."

This confession reminded Dr Strean of one of his own happiest memories. At 19, he had escaped briefly from the academic grind and worked outdoors selling ice cream. At college, he had written a composition about the happiest man he knew: it was the local garbage man. This had triggered Herbert Strean's own fantasies of doing his own thing and also being a garbage man.

Just like Saul, the young Herbert had resented his parents and other authority figures. So the rebel in him wanted to encourage Saul's dream. But since he believed that therapists should not tell clients what to do, he tried to remain neutral.

Gradually Saul came to accept Dr Strean's view that Charlie represented the little boy in Saul who secretly resented his parents' wishes, even though he felt obliged to follow them. At night Saul often dreamed that he was selling hot dogs at Disneyland. But then someone would destroy his stand, or he somehow lost it, and he would wake up depressed.

After nine months of therapy, Dr Strean suggested: "You want to give up medicine and sell hot dogs but you also want something to stop you."

Saul agreed: "If I gave up medicine, I'd be rejecting my parents."

After a year, Saul told Charlie about his dreams. Charlie surprised him by saying: "Daddy, you should do what makes you happy." Saul finally resolved to make his move. His wife thought he was crazy, but she was happy to escape the dreadful winters in New York.

Tearfully, Saul thanked Dr Strean: "You're the kind of parent I always wanted but never had." From California, Saul wrote about his happy days outdoors selling hot dogs. Charlie was also doing well.

The famous psychoanalyst Sigmund Freud wrote: "No psychoanalyst goes further than his own complexes and internal resistances permit." Dr Strean feels that he was able to help Saul only because he himself had dealt with similar conflicts during his own analysis.

But sceptics would say that this analyst actually encouraged his patient to do what he himself dared not do. Did Saul really jump or was he pushed?

(GB)

Dr Saul discovered true happiness in selling the humble hot dog. A happiness that no other fast food could bring!

TRUTH STRANGER THAN SOAP OPERA

It all started in 1974. He was Professor of Neurology at Rush University, Chicago. She came into hospital to have a small ulcer removed from one ear. She was a widow aged 64, a heavy smoker of 40 years who admitted she had smoker's cough. The intern confirmed that she had chronic bronchitis.

The anaesthetist ordered premedication, which included morphine. They gave her the morphine injection at 9.20 a.m. the next morning. When the porter came for her at 9.30 a.m., she was covered in a blanket and was half-asleep. The lift was as slow as usual, so she didn't reach the theatre suite until 9.45 a.m.. At 9.50 a.m., the theatre nurse came to check her in.

Red alarm.

Her lips were blue; she wasn't breathing. Then her heart stopped as well.

After the staff revived her, she got good heartbeats, but when they tried to stop ventilating her (pumping oxygen into her lungs), she could no longer breathe for herself. Again, they ventilated her, but once again they dare not stop. So they cancelled the surgery, kept her on a ventilator and settled her in the intensive care unit.

Then they asked the neurologist to assess whether she had suffered any brain damage. For half an hour, he talked to her. Because of the breathing tube in her windpipe, she had to write her answers. But her mind was very sharp; she told him all about her children and grandchildren, and even her favourite soap operas. He concluded that she had no brain damage.

Still, her own doctors just couldn't get her off the ventilator. Week after week, month after month, she stayed in hospital, reading, listening to music or knitting. She could do anything except talk or breathe for herself.

Then the hospital started buzzing with rumours of a court case.

People said she was suing; that with her bronchitis, the anaesthetist shouldn't have ordered morphine (which depresses breathing) for her. Once more they asked the neurologist to see her.

When he walked in, she was still on her ventilator but no longer sitting up, no longer reading. She was in a coma; no brainwaves; she was brain dead! The notes told him only that two days before she had suffered a cardiac arrest; they didn't say why or how.

When finally they had to take her off the ventilator, that was the end of her life, but not the end of the lawsuit. She had filed it, and her heirs kept it going. About two years after she died, the neurologist got a subpoena. The heirs were suing everyone. The hospital blamed the anaesthetist for ordering the morphine. He in turn blamed the hospital, especially the porter who had taken her to theatre in the beginning, for not seeing that she was not breathing. After six months, the case was settled for around US$1 million; one-third from the anaesthetist (or his insurer) and two-thirds from the hospital.

Only much later did the neurologist get the story from a defense lawyer.

The hospital administrators knew what had caused her fatal second episode. After her first episode, they had decided to teach all the porters resuscitation. But there was one porter who just wouldn't learn; he wouldn't even go to the classes. It was the same one who had taken her to theatre when she first arrested. The hospital felt they couldn't fire him — at least until the case was settled. Instead, they changed his job. They wouldn't let him move patients anymore; now he'd just deliver flowers, presents and television sets.

One day, the 64-year-old woman felt bored; she'd read too many books, listened to too much music. She asked for a television so she could watch her beloved soapies again. The porter came to the bed; she was snoozing, so he didn't need to wake her. He'd just make sure the set was working, leave it for her and go off to lunch.

All the power points were in use, so he unplugged one. The bedside light went out, so he plugged it back in and unplugged another. He tuned the television, turned it to *General Hospital*, and went on his way. Except of course that it was the ventilator he had disconnected.

It was only on the day before the porter was due to give evidence that the hospital settled.

(GB)

A GREAT WAY TO KEEP YOUR DATA SAFE

This story takes me back over ten years now. My wife Kitty started the whole computer thing. It was during what she lovingly called my "second midlife crisis", or when I was trying to start a writing career. I was thinking of an electric typewriter, but Kitty could see further ahead and suggested a word processor instead.

Our friend Phil took pity on my ignorance and helped me find a secondhand PC with the various extra bits. But now I needed a study large enough for all these bits with their cords and connections. It wasn't easy, but by juggling the kids' bedrooms, I finally managed. Only thing was, Kitty seemed quite tense by now.

At dinner parties, we no longer talked politics and cappuccino machines but RAM and bytes.

Even now Kitty still says I went overboard, I was neurotic, I was obsessional ... how wrong she was. I was simply being careful. How would you feel if a week's work vanished in a flash? At least she didn't argue about the surge board, but that's useless in a blackout. Obviously, an emergency generator was the way to go.

I joined Neighbourhood Watch. Etching my licence number on the computer, keyboard, monitor and printer was fine, but the floppy discs were harder. Kitty visits her parents each week, so I asked them to look after my third set of back-up floppies. They were fine about that, but Kitty got stroppy when I asked her to just update the floppies when she went over.

Then one day, we were planning to go out for dinner for my

birthday. I worked extra hard, finished an article about 4.30 p.m. and sent it from our post office (the poor man's fax machine).

On my return, I was surprised to see a delivery truck in our drive. Two burly men, breathing heavily, were battling with a huge steel-plate safe. Our daughter Jackie stood firmly barring the doorway. Her face was crimson and she kept shouting: "Take that thing back; it's all a mistake." To add to the uproar, her ghetto-blaster was blaring out "We shall overcome" at 100 decibels.

"There's no mistake," one of them snapped at me. "Just look at the docket." Sure enough, it had our name and address; it was all prepaid. "Just give me a minute. I'll ring your office and sort it out." But by now the office people had all knocked off.

What could I do? Without high explosives there was no way I could budge this monstrosity. Until we could talk to the office, we were stuck with it.

Jackie and I were trying to settle our frayed nerves with a strong coffee when Kitty bounced home from work. She edged her way past the safe in the hallway, jumped onto my lap and gave me a big hug: "It's come today, just as I asked them. I'm so glad for you. Happy birthday darling".

"Kitty, what on earth is going on? Did you order this thing?"

"Of course I did. You've been so worried about losing things from the computer, I had to help you. Don't you like your birthday present?"

"Just tell me one thing! Why do I need a safe three feet by three feet by three feet for floppies that are only five inches wide?"

"Don't be silly. It's not just for the floppies. This is your very own, very special safe. It's for the whole computer!"

(GB)

DOCTOR OSLER, THE JOKER

As you battle with the daily grind, do you ever yearn to do something childish just for the hell of it? Perhaps play a practical joke?

In 1884, the prestigious American journal *Medical News* carried a doctor's report of a house call:

> *I was sent for about 11 p.m., by a gentleman ... in a state of great perturbation ... At bedtime ... a noise in the coachman's room attracted his attention ... he discovered that the man was in bed with one of the maids. She screamed, he struggled, and they rolled out of bed together and made frantic efforts to get apart ... He was a big burly man ... and she was a small woman ... She was moaning and screaming, and seemed in great agony ... I found the man standing up and supporting the woman in his arms ... it was quite evident that his penis was tightly locked in her vagina, and any attempt to dislodge it was accompanied by much pain on the part of both ... I applied water, and then ice, but ineffectually, and at last sent for chloroform, a few whiffs of which sent the woman to sleep, relaxed the spasm, and relieved the captive penis, which was swollen, livid and in a state of semi-erection, which did not go down for several hours, and for days the organ was extremely sore. The woman recovered rapidly ...*
>
> *Yours Truly,*
> *Egerton Y. Davis*
> *Ex–US Army*

Over the years, journals and textbooks cited this case as a complication of vaginismus (contraction of the muscles around the vagina). But the account was pure fiction, written under an alias by Sir William Osler (1849–1919). Osler was a physician who became eminent in three countries, first at McGill University in Canada, later as founding professor of medicine at the Johns Hopkins University, and finally as Regius professor at Oxford.

His reputation, according to albert Lyons and Joseph Petrucell; in *Medicine: An Illustrated History*, is still secure as "a pragmatic, practicing physician who made outstanding contributions to medicine ... a writer of an encyclopaedic medical text which was a standard for generations ... the model of a cultured, articulate, insatiably curious, highly principled physician."

Does this sound like a man addicted to practical jokes? But the letter above was not Osler's only offbeat case report. Four years later, in 1888, in the *Canada Medical Record*, he described:

An interesting experience which I had [while on vacation by train] in the Northwest [of Canada] in 1886 ... [A friend] mentioned [that] two days before, a woman while in the water closet on the train, had given birth to a child which had dropped to the track and had been found alive some time after. I was so incredulous that he [the friend] ordered the conductor to stop the train at the station to which the woman had been taken ... I found mother and child in charge of the stationmaster's wife and obtained the following history:

"She was aged about 28, well developed, of medium size and had had two previous labours which were not difficult. She had expected her confinement in a week or ten days, and had got on the train to go and see her husband, who was working down the track. Having a slight diarrhoea she went to the water-closet, and while on the seat, labour pains came on and the child dropped from her. Hearing a noise and groaning, the conductor forced open the door and found the woman on the floor ... with just strength enough to tell him that the baby was somewhere on the track, and to ask him to stop the train, which was running at ... about 20 miles an hour. The baby was found alive off the side of the track a mile or more away, and with the mother was left at the station where I saw her. She lost a great deal of blood, and the placenta was not delivered for some hours [Osler made no mention of the umbilical cord]. I saw no reason to doubt the truthfulness of the woman's story, and the baby presented its own evidence in the form of a large bruise on the side of the head, another on the shoulder and a third on the right knee. It had probably

*fallen between the ties on the sand and clear of the rail, which
I found, on examination of the hole in the closet was quite
possible."*

This time the report bore, not an alias, but William Osler's own
name. Moreover, Osler held affidavits from the train conductor and
other witnesses. Not only did the baby survive, he lived for 74
years. But Osler was like the boy who cried wolf. How could his
colleagues know that this time he was for real? The editor of
Anomalies and Curiosities of Medicine hedged his bets. All he would
print was:

*There was recently a rumour, probably a newspaper
fabrication, that a woman while at stool in a railway car gave
birth to a child which was found alive on the track afterward.*

(GB)

Chapter 11

DISCOVERIES

WHEN LESS IS MORE: THE LEGACY OF SAMUEL HAHNEMANN

Let's face it, the principles of homoeopathy do sound odd. Here's a satirist's directions for homoeopathic resuscitation: "Lay one finger on the chest while blowing very gently in the general direction of the mouth."

But many present-day practitioners of homoeopathy are mainstream medicos. Moreover, some controlled trials suggest homoeopathy does work. Should we honour its founder, Dr Samuel Hahnemann (1755–1843) as an inspired trailblazer? Or did he just delude impressionable people with the power of suggestion?

Hahnemann was born into a poor family but was so gifted that supporters paid for him to study medicine in Leipzig and Vienna. He wrote on the treatment of venereal disease with mercurous oxide, translated many scientific works into German and researched the effects and detection of various poisons.

In his day, the mainstream medical treatments were bleeding, enemas, purging and induced vomiting. Many common medicines contained poisons like mercury or antimony. All in all, doctors may have killed more patients than their diseases did.

But Samuel Hahnemann was a rebel. He advocated good food, clean air and exercise and was among the few doctors to follow Pinel's enlightened treatment of mad people. Above all, he opposed the complex blunderbuss mixtures and medicines that doctors kept prescribing and chemists kept dispensing. The death of two of his own children from childhood illnesses made Hahnemann give up conventional practice and look for a better way. For the next few years, he supported himself by translating scientific works into German. One day, Hahnemann read about the remarkable benefits of Peruvian bark (later shown to contain quinine) for malaria. When he tried a little on himself, it produced symptoms like those of malaria.

That surprised him at first. Then he thought: instead of regarding fever as the *effect* of malaria, could not fever be the body's *response* to malaria? In that case, the bark's ability both to produce fever and to relieve malaria made sense. Did Hahnemann know that Hippocrates had said "like cures like [*similia similibus curantur*]"?

Over several years, Hahnemann tested many simple substances (derived from herbs, minerals or snake venom). First, he gave small amounts to healthy volunteers and noted their reactions. Later, when his own family or patients showed symptoms like those reactions, he treated them with a tiny amount of that same substance. Onions made healthy people's eyes water and noses run, so Hahnemann found onions good for the common cold. Arsenic produced the stomach pains, vomiting and diarrhoea of dysentery, so he decided arsenic was good for dysentery.

But Hahnemann used the smallest possible doses of drugs. According to his theory of "potentisation", medicines *gain* in potency by being diluted, as long as vigorous rhythmic shaking or pounding accompanies the dilution.

How did he come to use such dilute mixtures? How did he convince himself, his followers and his patients (surely they were not all his relatives!) that umpteen dilutions were not only safer, but also *more effective*? By the time Hahnemann was through with his endless diluting, his mixtures might not have even contained one atom of his original active drug. On this issue, his critics have had a ball. One doctor said that a dose which an English practitioner would give to a suckling baby would, in the hands of Hahnemann's disciples, cure everyone in the whole solar system!

He taught that all diseases resulted either from conventional medical treatment or from one of three conditions: psora (the itch), the skin disease sycosis (papillomatous warts) or syphilis. Indeed, most chronic diseases were caused by psora being driven inward. Hahnemann called all other (non-homoeopathic) systems of medicine fraudulent. He had the fanaticism and arrogance of many crusaders: "He who does not walk on exactly the same line with me, who diverges, if it be but the breath of a straw to the right or left, is an apostate and a traitor ... "

But he did seem to get results. Chemists denounced him for using only one drug at a time and, even worse, for dispensing all his own medicines. Unable to handle his popularity with patients, jealous physicians had Hahnemann run out of Germany. In 1835, he remarried and set up in Paris, where his practice flourished and he became a millionaire. He remained active until his death at the then remarkable age of 88. Since then, homoeopathy has spread widely, and Hahnemann's name and teachings still live on in some of the world's most prominent medical schools and hospitals.

To return to our original question: was he a quack or a genius? It all depends on whom we ask. The famous American medico Oliver Wendell Holmes (1809–1894) called homoeopathy "a mingled mass of perverse ingenuity, of tinsel erudition, of imbecile credulity and of artful misrepresentation".

Holmes may be too harsh. Even without necessarily believing that Hahnemann did any objective good, we could accept that he did get better results than those practising orthodox polypharmacy, bleeding, puking and purging. Even critics agree that homoeopathy must at least have been pretty harmless; and medical schools still teach students that, above all, they must do no harm.

But there is evidence that Hahnemann and homoeopathy have done better than this. In 1813, when an epidemic of typhoid (or perhaps typhus) swept through Leipzig, Hahnemann treated 180 patients, of whom only one died! Later, during London's cholera epidemic of 1854, the death rate at the Homoeopathic Hospital was said to be only 16%, compared to 52% elsewhere.

Moreover, Geoffrey Watts wrote "a review of the literature in 1991 identified over 100 published controlled trials of homoeopathy, and found that while its case could not be taken as proved, the balance of evidence lay in favour of accepting that it works."

Homoeopathy alive and well in 1999

Dr Eric Asher and Dr Nick Goodman are associates in a busy general practice in the Sydney suburb of Lindfield. Both graduated from Sydney University in the 1970s and have postgraduate medical qualifications. Both also have British experience and qualifications in homoeopathy.

Both integrate homoeopathy with their mainstream medical training and use the former as a cheap, simple approach, especially for their less ill patients. Dr Asher's examples: "Viral illnesses, coughs, colds, sore throat, childhood fevers, bee stings and hangovers. The herbal remedy Arnica is strikingly effective for sprains and bruises, even those following surgery or childbirth. A homoeopathic treatment often works within minutes, while the patient is still with me. But if the response is not so clear, I may also write a conventional script to be filled later if necessary. Only about one-third of patients actually use these scripts."

Dr Goodman works similarly. "Homoeopathy does not work all the time, but neither do antibiotics. When homoeopathy does work in infection, it stimulates the immune system, is cheaper than an antibiotic and does not stimulate germs to become resistant to drugs."

The million-dollar question: can we explain why a super-diluted mixture should work?

American researchers have photographed homoeopathic solutions. They found that the original active ingredient or solute changes the water or solvent. As one dilutes and redilutes the solution, even at room temperature, specific water crystals (visible under the electron microscope) form and these crystals carry the specific code of the active solute. Different solutes trigger different crystals, and these crystals may be therapeutic even when the solute itself is too diluted to be effective.

This fundamental research is still in progress. Clinical trials of homoeopathy appear in the *Lancet,* the *British Medical Journal* and the *British Journal of Homoeopathy.*

Drs Asher and Goodman run courses on homoeopathy and general health care for their patients. These aim to inform people and so make them feel more self-reliant and in control.

(GB)

MEDICINAL PLANTS: A KEY LINK WITH OUR PAST

Medicinal drugs all come from chemists working in factories. Right? Wrong. About three-quarters of the world's population depend on traditional medicine, which in turn depends largely on drugs from plants.

The use of medicinal plants is as old as humanity itself. Our ancestors were not just hunter-gatherers; they were also practical botanists. Juice from the opium poppy was popular at least 6000 years ago. Chinese records of medicinal plants go back at least 4000 years. Egyptians used ergot and about 800 other drugs. The *Rig-Veda*, the oldest sacred book of Hinduism, lists 700 medicinal plants.

Even in "advanced" countries, medicinal plants are of far more than historical interest. In Australia, though we use many synthetic drugs, we still depend on the plant world as well.

How come? Our range of synthetic drugs is far from complete. We still lack effective drugs for many diseases and some germs become resistant to our antibiotics. Plants make far more chemicals than we humans have ever devised. So we turn to plants to find new drugs, less toxic drugs or cheaper drugs. Perhaps one quarter of our synthetic drugs were first isolated from plants.

In the 1950s, researchers studied the Madagascar periwinkle, a folk-remedy for diabetes. They derived vincristine and vinblastine, now used for Hodgkin's disease and childhood leukaemia.

The Second World War stimulated Australians to use the Australian native *Duboisia* to extract hyoscine to prevent motion sickness in our armed forces. We still extract some hyoscine in this way, but most of the leaves go overseas for processing. We used to make digoxin (for heart failure) from our own foxglove, but South African plants have replaced this.

Manufacturers still derive morphine and codeine from the Tasmanian opium poppy. A potent drug for cancer of the uterus and ovary is taxol, which is derived from the yew tree.

Plants, especially those in rainforests (including our own in North Queensland), offer the world an enormous, still largely untapped laboratory. Of about 250 000 known flowering plants, we have analysed the medical potential of only 5000.

So the forests still contain many useful plants that we should check out. That's great news for us, our children and grandchildren. But the bad news is we have already lost about half of the world's tropical rainforests. What to do? Perhaps go green.

(GB)

FRENCH FREEZE OUT FOREIGN FEMINIST PHYSICIST

At least she is in Paris. But not the Paris of anyone's dreams! An abandoned shed: just a few boards thrown together; bitumen floor; a cracked skylight that lets in the rain. It was a dissecting room but is no longer fit even for corpses, stifling in summer and freezing in winter.

Outside in the courtyard, the woman battles with a huge, heavy iron rod, stirring a large cauldron brimming with a boiling, volatile liquid. Occasionally she stops to pour the liquid into a jar, which she drags into the shack. When all the jars are full, she joins the man inside.

Marie and Pierre Curie toiled like this for years on end. Born in 1867, Marie (Manya) Sklodovska was the youngest daughter of well-educated Polish patriots. Marie's mother was dying of tuberculosis. Only much later did Marie understand why her mother never kissed her. Russia ruled their part of Poland. Her father, a physics professor, was demoted when the Russians found him teaching his Polish students their own language. He lost his life savings just when Marie was about to enrol at the Sorbonne in Paris. So Marie first worked for five years to support her elder sister, Bronya, while the latter studied medicine. After graduating, Bronya in turn helped Marie.

In 1893, living on tea, bread and butter, Marie topped her class and got her master's degree. For her wedding to another physicist, Pierre Curie, she chose a dark suit that would not show the stains of lab work.

It was an exciting time to be working in physics. Soon after Roentgen's discovery of X-rays, Antoine Becquerel found that uranium emitted mysterious rays that penetrated solids. For her doctorate, Marie explored other sources of this radiation, which she herself named radioactivity. She found that the radiation from pitchblende (the main ore of uranium, mined in Bohemia) far exceeded that expected from its content of uranium. Hence, she concluded, pitchblende must also contain another very radioactive, but still unknown, element.

In fact, the Curies discovered not one but two new radioactive elements (both breakdown products of uranium): polonium (named after Poland) and radium. Radium was one million times more radioactive than its parent uranium! It took several years to realise that the Curies' gamma rays were identical to Roentgen's X-rays.

But to isolate radium was another matter. So poisoned was the atmosphere in the derelict dissecting room that the Curies' notebooks are even now still dangerously radioactive. Pierre's hands became so scarred that he could not even knot his tie. Later he developed cancer. In 1901, the Curies lent their friend Becquerel a tube containing radium. After carrying it in his pocket for only six hours, Becquerel found it had burned his skin, just like a burn from X-rays. This triggered interest in radium's possible medical effects. Finally, in 1902, after four years of toil, from eight tonnes of pitchblende, the Curies isolated one tenth of a gram of radium! Marie and Pierre shared the 1903 Nobel Prize in physics with Becquerel. The world cheered.

A farmer wanted to put radium into his chicken-feed so his hens would lay hard-boiled eggs. An academic at Columbia said radium-fertilised soil would yield more and tastier crops. Reportedly the Shah of Persia came all the way to France to see radium. When he saw the jar of crystals glowing in the dark, he got so excited that he jumped up, knocked over the table and broke the jar. Marie kept cool: she dissolved the carpet in acid and recovered her radium.

Pierre and Marie could have taken out patents and got rich, but instead made public their findings for others to apply. Life was looking up when, in 1906, Pierre died after being run over by a lumber wagon. Marie worked on despite this bereavement, the first symptoms of radiation sickness and the insults of the French. As a foreigner and as a woman, she never won over the French authorities. Very reluctantly, they awarded her Pierre's professorship and asked her to give his lectures at the Sorbonne. Marie was the first woman to do so.

She found that the luminous substance she and Pierre had called radium was actually a salt of radium. But Marie finally did isolate radium itself. She produced the first international standard of radium. The *curie* became a unit of measurement of radioactivity. In 1911, she won her second Nobel Prize (this time for chemistry), but still the French Academy of Science would not admit the "foreign woman".

Concrete-encased chook shed

WARNING
HIGH CHICKEN RADIATION

HARD-BOILED EGGS 4 SALE

The "brilliant" farmer thought that feeding radium
to his chickens would produce hard-boiled eggs.
And, as an added bonus, they had a nice green
glow, perfect for illuminating the breakfast table!

When the First World War came, she invented mobile X-ray
vans ("little Curies") to locate pieces of shrapnel in wounded troops.
She trained 150 people as X-ray technicians, raised money to equip
the vans, and she herself drove one. By the end of the war, her 20
vans and 200 X-ray posts had examined over a million men.

In 1921, Marie went to the USA, where President Warren G.
Harding presented her with one whole gram of radium (worth
$100 000) for medical use in Europe. Her older daughter, Irene,
worked with Marie on the medical applications of radioactivity.
Warsaw set up a Radium Institute, with Marie's sister Bronia
as director.

It was radium, the element to which she had devoted her life,
that killed Marie. When she died in 1934 of leukaemia, her body

joined Pierre's at a cemetery. But over 60 years later, in 1995, the French finally moved her remains to the pantheon. She is the first woman to be so honoured. Finally now, Marie Curie rests with the foremost sons of France.

(GB)

Chapter 12

MAYHEM AND MURDER

THE BLOODING OF GENETIC FINGERPRINTING

Six miles from the county seat of Leicester, England, is the village of Narborough. It was here that a passer-by found the body of Lynda Mann in November 1983. Her scarf was wrapped around her neck and knotted at the back. The autopsy showed that she had been strangled. There had been attempted intercourse and premature ejaculation. On antigen testing the semen showed group A secretor substance.

"Village of fear" was how one paper described Narborough that winter. A murder squad of 150 started the investigation. They had plenty of leads. Over 100 potential suspects took blood tests, but by August 1984 there were no suspects left. In desperation, Lynda's parents went to a medium who predicted that the murderer would strike again.

At nearby Leicester University, a young geneticist, Dr Alec Jeffreys, was developing the technique to be known as genetic fingerprinting.

In August 1986, Dawn Ashworth, another 15-year-old schoolgirl, was found strangled. She had also suffered a brutal sex attack. A businessman offered £15 000 for information leading to a conviction. Police soon arrested a 17-year-old porter at the local psychiatric hospital. He was known to molest girls and had talked of Dawn's body being found before it had become common knowledge. His story changed with every questioning, but he did confess, though only to the second murder.

Dr Jeffreys found that the porter's blood did not match the three-year-old semen sample from Lynda's body. It took another week to test Dawn Ashworth's sample. The tests showed that the porter had not committed either murder, but that the same man had murdered both girls.

Many of the 1800 leads came anonymously. One tip was about a baker, Colin Pitchfork, who had no alibi and had a record for flashing. Perhaps because Colin had not moved to the village until after the second death, police did not follow this up. Nor did Colin's wife tell anyone what really turned him on: he liked her to wear long white socks, like a schoolgirl.

Now police set up voluntary blood and saliva testing. They targeted males who had been between 13 and 30 when Lynda Mann was killed. Why they excluded older men is not clear.

Colin felt worried when he saw the headline "Blood tests for 2000 in killer hunt". After he got the second letter, his wife pushed him to go, but he said that his old record of flashing would get him into trouble. Quietly he asked several workmates to take the test in his name. No luck. His last chance was young Ian Kelly. At a bakery Christmas party, Colin had propositioned Ian's wife, who did not tell Ian. Later she wished she had. Finally, Ian agreed to Colin's ruse. Colin stuck Ian's photo in his own passport and drove him to the blooding. Ian carried it off.

By May 1987, over 3000 males (98% of those invited) had given samples but the laboratories were way behind.

On the anniversary of Dawn's death, a plainclothes man with a video camera was watching the murder site, but saw nothing.

Dawn's parents were at the time visiting relatives in Australia; they timed the dateline crossing so as to make the anniversary disappear: an extreme case of denial.

In August 1987, Ian Kelly was drinking at a pub with a bakery manager and other workers. When the talk turned to Colin Pitchfork, Ian let slip that he had taken the blood test for Colin. The manager wanted to go to the police, but the others didn't want to make waves. The publican's son was a constable, but he was on holiday. It was six weeks before the manager finally talked to him.

Police compared Colin's own signature with Ian's forgery from the blooding; they were quite different. The police arrested Ian, who admitted to the impersonation. Colin told police his whole life story, the flashings as well as the murders and rapes. After killing Dawn, he had gone home and baked a cake: he showed no remorse. A psychiatrist called him a psychosexual psychopath.

He got a double life sentence for the murders, 10 years for each rape, three years each for earlier assaults and three years for the impersonation. But this sentence means little, since he would serve the terms concurrently and there was no minimum term.

The police were outraged; the parents of Lynda and Dawn wanted hanging brought back; 96% of locals who took a poll agreed. The superintendent in charge of the second case said God must have had a hand in this DNA business, which had not only led to the release of the porter but also to the conviction of Colin Pitchfork.

(GB)

MEDICAL PATRIOT OR AGENT OF DEATH?

By 1942, the Nazis had invaded France and installed the collaborationist Vichy Government. Jews in France had every reason to fear persecution from both the occupying Germans and Marshal Petain's puppet government. Why wait for the noose around their necks to tighten? Why not try to escape? Many Jews turned to the Resistance for help.

A Mr and Mrs Wolff consulted a Dr Eugéne who ran his own underground escape system (*Fly Tox*). He told them to take as much money as they wanted, but to remove all labels from their clothes. While they were in a safe house waiting for their new papers, he would give them inoculations for the journey. The Knellers paid Dr Eugéne 1 500 000 francs. Dr Braunberger paid 1 000 000 francs to escape via Spain. There were other clients as well: Paulette la Chinoise planned to open a brothel the minute she reached South America.

Dr Eugéne kept his word — his clients all disappeared from France.

But one family pulled out at the last minute; the wife was a doctor. No problem; he just refunded their money.

By now the Germans were closing in; soon they arrested "Dr Eugéne" as Dr Marcel Petiot, a Paris GP. Though it is reported that they interrogated and tortured him for eight months, the patriotic Dr Petiot would not name his Resistance friends in the escape network. Finally, in December 1943, Petiot was released and was soon practising again.

A few months later, neighbours complained about the foul, greasy smoke billowing from the chimney of a large house belonging to Petiot. There was no answer to their knocking, and all the doors were locked. By now the chimney was alight! They called the police, who forced their way in. The first man to enter the house came out vomiting. In the basement, a huge furnace was blazing; from the open door, there drooped a human hand. On the stairs were skulls, limbs, hands, feet, other bones and masses of flesh.

A man came forth, saying he was the owner's brother, head of a Resistance group. Moreover, he said, the bodies belonged to Germans and to French collaborators. Petiot always told people what they wanted to hear; the patriotic French officials advised him to vanish.

Only then did they find the tiny triangular room without windows; the single door was soundproofed and opened only from the outside; it had a spyhole. On one wall were eight iron rings. In the garage, from a huge pile of quicklime, they sifted many human parts. A manure pit in a stable held more bodies. Medical experts spent months sorting the human parts. But they could piece together only ten bodies, none of which they could identify.

Petiot vanished for eight months.

The Police Commissionaire leaked a story to the papers accusing Petiot of having worn a German uniform and having worked against the French Resistance. The super-patriot Petiot took the bait and wrote a furious denial. This letter suggested that he was still in Paris, serving with the French forces. Soon the police tracked Petiot down and arrested him.

Finally, in March 1946, he faced 27 counts of murder. Though the Nuremberg war trials were on, in Paris it was Petiot who got the headlines. Details of his past came out. In the First World War, he had dealt in stolen drugs. He avoided the front by shooting himself in the foot and by throwing convulsions. About 1918, he got a disability pension and entered a mental hospital. But somehow he became a medical student and even graduated in 1921.

Later a pregnant mistress disappeared very conveniently. In 1930, there were rumours of robbery and murder, perhaps even a second murder to cover up the first; also fraud, illegal abortions and supplying narcotics to addicts. Remarkably, no medical authorities had ever questioned his fitness to practise medicine!

But Petiot was amazingly popular. Many patients told police the good doctor would often cycle miles to see a sick child and would not charge poor patients. Locals had twice elected him mayor.

Still the accusations of murder remained. But no-one had seen any victims enter his house; none of the corpses had been positively named, and pathologists could not show the cause of death. Petiot admitted to killing 63 people, but insisted it was all in the name of the Resistance and for "the glory of France". The courtroom clapped when his lawyer proudly declared: "He brought down his enemies, our enemies." Had Petiot not resisted torture at the hands of the Gestapo?

But the prosecution called over 80 witnesses: Resistance heroes had never heard of *Fly Tox* or of Petiot. If his clients had escaped, why had none of them ever contacted their relatives in France? How could the jurors ignore the immense pile of victims' suitcases in the courtroom?

After midnight on the 16th day came the verdict: Petiot was guilty of 24 murders and would face the guillotine. Only then did Doctor Death's mask fall as he screamed, "I must be avenged!"

Had he not chosen the path of murder and fraud, Dr Marcel Petiot might well have been a hero.

(GB)

THE RISE AND FALL OF
DR RODERIGO LOPEZ

Doctors should stick to medicine. Political intrigue is dangerous; even to be accused of intrigue ...

Nineteen ninety-four saw the 400th anniversary of the execution in London of Dr Roderigo Lopez, physician to Queen Elizabeth I. Lopez was born around 1520 in Portugal. The Inquisition there forced him, like many other Jews, to convert to Christianity and to leave Portugal. He probably studied medicine in Spain.

Finally, Lopez settled in England. Soon he became senior doctor at St Bartholomew's Hospital. By 1586, he was physician, companion and confidant to the Queen herself. Being fluent in Portuguese, Spanish, Dutch, Latin, Hebrew and English, he often translated documents of state. When she granted him the monopoly to import the herbs aniseed and sumach from America, many Englishmen were jealous.

To understand his life and especially his death, we must remember the prolonged conflict between the Protestant Queen of England, Elizabeth I, and the Catholic King of Spain, Philip II.

**Dr Lopez: born 1520, physician to Queen Elizabeth I,
a Jew, a convert to Christianity, forced to leave
his home of Portugal and later in life ... executed!
All in all, a pretty full résumé you might say!**

In 1588, Sir Francis Drake defeated the Spanish Armada which had been sent to conquer England. Thereafter, some, but not all, English politicians wanted to make peace with Spain.

Much of the intrigue at both the English and the Spanish courts surrounded Don Antonio, the illegitimate son of the heir to the Portuguese throne. Philip of Spain forced him to flee to England, where the hawkish Robert Devereux, Earl of Essex, supported him as a tool against Spain.

Among the prominent patients who flocked to Lopez was Secretary of State Sir Francis Walsingham, who also recruited Lopez to spy for England.

In 1593, a Spanish agent tried to bribe Lopez with a diamond and ruby ring to poison Don Antonio. Lopez reported the plot to Elizabeth who pooh-poohed the story. Later, Spanish agents offered

Lopez 50 000 crowns, which he refused. Essex got wind of this and also suspected (or claimed to suspect) a plot to poison Elizabeth herself. Under torture, the Spanish agents tried to save themselves by implicating Lopez in such a plot. In January 1594, Essex arrested Lopez himself and threatened to torture him as well. Eventually, the terrified old man made a so-called "confession".

Now the Spaniards and Lopez faced a harrowing cross-examination at London's Guildhall by a commission that included Essex. Lopez again proclaimed his innocence but was convicted. Essex declared: "I have discovered a most dangerous and desperate treason. The point of conspiracy was Her Majesty's death. The executioner should have been Dr Lopez; the manner poison." By now Walsingham, who would have protected him, was dead. Both anti-Spanish and anti-Semitic feelings were directed against Lopez; Englishmen published drawings of him mixing poisons.

The penalty for conspiring to poison the Queen was death, but Elizabeth, still convinced that Lopez was innocent, refused to sign his death warrant. Essex had Lopez moved from the Tower to Southwark Prison, where death warrants did not need her signature.

On 7 June 1594, Lopez and the Spanish agents were dragged through the streets to Tyburn (now the site of Speaker's Corner in Hyde Park). There he declared: "[I love] the Queen as much as Jesus Christ." The mob jeered. The men were hanged, drawn and quartered: first hung from the scaffold, next cut down while still alive, then castrated, disembowelled and chopped into quarters.

No-one emerged with credit from this brutal story. Political intrigue and religious persecution are to blame for these deaths.

Elizabeth restored Lopez's property to his widow. The ring that Lopez had offered her she wore in her belt for the rest of her life. It is reported that documents found much later in Spain cleared Lopez. Essex's accusation that Lopez planned to poison his patron, Queen Elizabeth, seems absurd. Having been forced by the Inquisition to flee from his native Portugal, why would he want to kill his patron and protector?

Lopez never knew that the stubbornness and ambition of Essex finally lost him Elizabeth's favour and led Essex himself to be executed for treason in 1601.

A revival of Christopher Marlowe's anti-Semitic play *The Jew of Malta* at the time of the trial may have fanned English hatred of Jews. Was there also a link between Lopez and the character of the Jew Shylock in Shakespeare's *Merchant of Venice*, which played in London two years after Lopez died?

Even today, visitors to the Guildhall may still notice a sign headed "Famous Trials Held in This Hall". The names include Lady Jane Grey and Archbishop Thomas Cranmer and the sign notes: "Dr Roderigo Lopez, physician to Queen Elizabeth, was tried for High Treason, and was executed at Tyburn in 1594".

(GB)

ADDICTS MAKE DEADLY DOCTORS

The son of an Anglican clergyman, Dr George Henry Lamson was born in 1849. He won decorations as a volunteer army-surgeon in Serbia and Romania. Yet only a few years later, he was convicted of perhaps the most diabolical murder of the century. How could a man fall so far so quickly?

In the Balkans, Lamson suffered war wounds and become addicted to morphine and perhaps also to aconitine, the active principle of the wolfsbane plant. Mainstream doctors used aconitine in liniments, though Lamson prescribed it indiscriminately for most ailments.

Back in England, the young doctor married Miss Kate John. On marrying, she came into a small fortune, but before the Married Woman's Property Act, her money passed into the control of her husband. Kate was as devoted and submissive as "a feudal Japanese wife". She had two brothers, Herbert and Percy, and one sister.

When Herbert died suddenly, Kate inherited £700, with which Lamson bought a practice in Bournemouth. But he passed too many dud cheques and bailiffs called more often than patients.

He also had paranoid delusions. Why else would he fire a revolver from his bedroom window, swearing that Turks were about to attack him? The local medical society expelled him. He had to sell their home and pawn his watch and even his surgical instruments.

But if his remaining brother-in-law, Percy, were to die unmarried before turning 21, Kate would inherit £1500.

Percy was a boarder at Blenheim School, Wimbledon. Despite the paraplegia that bound him to a wheelchair, he was a cheerful boy. In mid-1881, when Percy came to spend his holidays with his sister, Lamson probably slipped him a dose of aconitine. This gave Percy nasty heartburn, but he recovered.

On the evening of 3 December, gaunt and excitable, Lamson called at the school to see Percy. With him he brought sweets and a currant-studded, almond-topped Dundee fruit and nut cake, probably already sliced. Percy, the headmaster and Lamson himself all had cake and sweets. Soon Lamson left, caught the boat train from Victoria and crossed to France.

Within 20 minutes, Percy complained of heartburn. Between frequent, copious vomits, he suffered such agonies that he had to be held down. Two doctors came, but nothing helped. After four hours of agony, poor Percy died. The doctors collected a sample of his vomitus. Though the autopsy report was pretty normal, everything pointed to poisoning and to Lamson as the poisoner.

On 8 December, Lamson, bold as brass, returned to London and went straight to Scotland Yard to "clear his name". In his luggage, police found a book on vegetable poisons. The Home Secretary ordered a further autopsy. Pathologist Dr Stevenson found gross irritation of the stomach, but also a little aconitine in a raisin from Percy's gut.

The Old Bailey was packed as the trial started on 9 March 1882. Lamson pleaded not guilty. Dr Stevenson believed that Percy had been poisoned. Lamson could have somehow marked Percy's slice of cake, pulled out a few raisins, filled them with aconitine and replaced them. The crown had a strong case, except that there was then no chemical test to detect vegetable poisons. Forensic experts had to taste extracts from various organs after death! Dr Stevenson said he had learned to identify 50 substances by taste! He testified that Percy's bowel contents had yielded a distinct substance. Moreover, injecting mice with extracts from Percy's organs had produced effects identical to aconitine.

The jury took only half an hour to find Lamson guilty. But fashionable women bombarded him with flowers and gifts. His defenders claimed that morphine addiction had destroyed Lamson's sense of ethics. Even the American President, Chester Alan Arthur, intervened on his behalf. All in vain.

Waiting in Wandsworth Prison, Lamson found the cold-turkey withdrawal from morphine devastating. Finally, he confessed to Percy's murder, hoping that his career would give "an awful warning to others, similarly tempted and assailed, seeing to what fearful consequences morphine addiction has led".

Even as the hangman was pulling the lever, Lamson begged for time to hear just one more prayer.

(GB)

Chapter 13

Midwives, women and babies

A history of midwifery

Of the three main streams within the discipline of medicine, namely pure medicine, surgery and obstetrics, that of obstetrics is very much a Johnny-come-lately. Physicians look back on their lineage as being an intrinsic, essential and perhaps glorious part on life's grand plan for society, with clinical medicine at the forefront. Surgeons take bogus pride in their humble beginnings and early association with barbers but are quick to point out that Henry VIII was their first mentor. But obstetricians have not completely thrown off the ignoble feeling of being male midwives and a poor relation of the other disciplines. Only in this century have they been formed into a Royal College, although their battle to wrest power from the neighbourhood handy woman took place about 200 years ago. At this safe distance the battle makes salutary reading, so let's look at it.

The doctor whose name is the first to be associated with the rationalisation of childbirth is William Smellie, born in 1697 in Lanark, Scotland, and a graduate of Glasgow University. There had, of course, been many male accoucheurs before him, but it was Smellie who published the original primer on the subject. It was called *A Treatise on the Theory and Practice of Midwifery* and appeared in 1752. Smellie had a stroke of luck right at the beginning, for he was assisted in this literary task by one of the foremost writers of the day, Tobias Smollet, himself a doctor and author of *Roderick Random*. Doubtless, the partnership made the textbook more readable, so gaining it a wider audience. At the time of publication, Smellie was 55 years old, and, by dint of much practice in the Scottish countryside, as well as in Paris and in Wardour Street in Soho, London, his knowledge on the subject was encyclopaedic. Thus armed he proceeded to correct many of the fallacies which had been passed down over the years from one dubious practitioner to another.

For instance, it had always been maintained that the foetus lay with its head uppermost in the uterus until the seventh month, when, due to it becoming top heavy, the baby turned turtle, ready to escape by, so to speak, crawling out on its hands and knees. True enough it often has the head high in the womb early in pregnancy, but crawling out was surely a bit fanciful. He also noted that the head did not come straight down through the pelvis like a diver from the high board, but took the line of least resistance and progressed with a kind of slow screwing motion through the passages. He worked out his theories by observation and careful measurements of the female pelvis.

Of course, Smellie had centuries of prejudice and misinformation to overcome, starting with Hippocrates (460–357 BC), the Father of Medicine, who, amongst other gems, thought that labour was started by the efforts of the baby struggling to get out when it had gathered enough strength to do so. This picture of the infant clawing its way to the outside world persisted for almost 2000 years. It was even shared by the great William Harvey a hundred years

**Dr Smellie, although trying to prevent conception,
sounded just plain stupid with his post-coital
pillow talk, which was often misunderstood completely.**

before Smellie, and who wrote, "the foetus promotes his own delivery by his efforts, like a butterfly emerging from a chrysalis". Maybe a poetic appraisal, but, in passing, note the parsing of "foetus". I wonder if he considered that female infants had such a kinetically charged arrival.

But let us go back to Hippocrates in the fourth century BC. He thought, not unreasonably for the era, that any presentation other than head first was dangerous. After this good start he fluffed it by recommending that to ensure cephalic, or head, presentation the unfortunate lady should be strapped to a board which was then put upright and repeatedly struck on the ground.

The next great name to emerge in the field of obstetrics was that of the Greek Soranus, from Ephesus. This is now a splendid ruined town in modern Turkey. Soranus lived in the second century AD and like many of his countrymen sought his fortune by setting up practice in Rome. He was a sound clinician and acute observer, relying on simple rules to explain disease, and attributing illness to an adverse state in the "internal pores".

His monumental work, *On Midwifery and the Diseases of Women*, contains descriptions of contraceptive methods and delivery of babies feet first. He distinguished between contraceptives and abortifacients, considering the former more desirable. Unfortunately, his advice as to how to avoid conception would seem to lack the ingredients for success, for he advocates holding the breath, coughing, jumping and sneezing after coitus in order to expel the semen. Regrettably, he does not give success rates, and I suppose that with the advent of more reliable methods such as the Pill, the moment to carry out clinical trials has passed. But there is no doubt that if assiduously followed the method would at least add a new dimension to a couple's sex life.

Although Soranus wrote a great deal on the subject of obstetrics and gynaecology, incredibly his chief manuscript was not found until 1838 when it turned up in the Vatican library. This is a pity because the skill he taught in podalic version, or delivery feet first, was forgotten for 1400 years until Ambrose Pare (1510–1590) in Paris redescribed it, thinking he had discovered it for the first time. Soranus advocated embryotomy, or sacrificing the child by dismembering the body, in a case of difficult labour. This barbaric practice survived until comparatively modern times when anaesthetics thankfully changed the outlook.

The Middle Ages were a thin time for all medical progress, including obstetrics. But at least one odd piece of apparatus was devised then and came into constant use — the birthing stool. Such an item of furniture formed part of the dowry of every bride in many countries until the seventeenth century. It was a wooden affair with a straight high back and a seat constructed in the shape of a horseshoe to leave a large hole in the middle. Only in the following century was delivery in bed regarded as the norm. This preferred lying position is still disputed by some today.

Eventually, medicine in general emerged from the old Greco-

Roman ideas of Galen and others with the groundbreaking discovery by William Harvey in the seventeenth century of the circulatory system. As well as this fundamental work in physiology, Harvey was also a distinguished obstetrician. Towards the end of his life he wrote *De Generatione*, about 30 pages of which deal with his experience in the delivery room. He was a great believer in the "power of Nature", and deplored the meddlesome efforts of some midwives to hasten the birth process. "Nature herself must be our adviser," he wrote. "The path she chalks must be our walk."

In another section he was more pointed in his remarks when he expostulated, "the young, giddy and officious midwives do persuade poor women to use their birth stool before the time and do bring them in danger of their lives". This kind of gratuitous talk did not endear him or his fellow doctors to the midwives of the day. I am sure that they did not consider themselves officious, nor probably young in the sense of inexperienced, and most certainly not giddy. I suppose what Harvey lacked in tact, he made up in courage. Other medical men added their weight to the castigations. Peter Willoughby from Derby wrote of the "incessant and violent interferences of ignorant midwives" — fighting talk to an ancient and well-entrenched profession.

Of course, many of the patients refused to be attended by a man. Decorum was at stake. Willoughby instructed his daughter in the art and she would call him in if she felt complications were at hand. So as not to affront the patient by his masculine presence, it was his practice to creep unseen into the delivery room on his hands and knees.

The Royal College of Physicians would have no part in the training of midwives, especially in this deteriorating obstetrically political climate. But at least the Church of England accepted responsibility for their control, and in a fit of ecclesiastical whimsy the Archbishop of Canterbury, no less, granted them licences. Perhaps rather prudently, the Church did not dabble in training or education, just a licence. The appropriate piece of paper given to one Eleanor Pead giving her permission to ply her trade reads in part: "I promise I will be ready to help poor as well as rich women in labour; I will not suffer any other body's child to be brought to the place of a natural child; I will not use sorcery or incantations; I will not destroy, cut or pull off the head of any child, and I will, in time of necessity, baptise the child with the accustomed words, using pure and clean water."

That clause about substituting another child may have been included following the remarkable incident concerning Mary of Modena, wife of James II. She delivered a male child on 9 June 1688, but her jealous stepdaughters, Mary and Anne, both of whom were destined to become Queens of England, doubted the authenticity of the pregnancy. They claimed that a baby boy had been smuggled into the delivery suite in a warming pan, then touted as the royal issue. There is no question that a male child was produced. He became James Francis Edward Stuart, better known later as the Old Pretender. On account of his Catholic background he was disinherited by Anne so he never became king.

Although the handy women undertook not to pull off children's heads, the medical profession was not convinced that such instructions would be followed. Perhaps they feared the pious incantations had merely become *sotte voce* and were being turned onto them. Feelings were running high, as illustrated in 1769 when William Buchan, an obstetrician distinguished enough to be buried in Westminster Abbey, wrote, "Few women think of following the employment of midwife unless they are reduced to the necessity of doing it for bread. Not one in a hundred of them has any education or proper knowledge of her business."

It must have taken a brave man to write such things, and there is no doubt there was a great deal of rivalry between midwives and obstetricians over several hundred years and that it was at its most astringent in the eighteenth century. Indeed, in England it was only as recently as 1902 that the Midwives Act actually defined "midwife". The animosity generated was not calmed by the use of two operative procedures to which the doctors had access but the midwives did not — obstetrical forceps and Caesarean section.

Fact and fiction are hopelessly entangled in the early history of

Caesarean section. Macbeth believed he could not be killed by "one of woman born", only to have his confidence undermined by Macduff's revelation that he was, "from his mother's womb untimely ript". Perhaps from the word "untimely" we can infer that this was a postmortem operation, as I am sure all early such procedures took place after the death of the mother, a not altogether uncommon happening.

Popular belief is that the operation is so-called because Julius Caesar was delivered this way. We know that his mother, Aurelia, was alive when he invaded Britain, so it was certainly not a postmortem operation in his case. Even if at the time it had been possible to do it in life, it is still unlikely as, by contemporary custom, the doctor would have been a Greek slave, and he would hardly have dared to even suggest such a risky procedure on the Emperor's wife.

A more likely explanation for the name is that in 715 BC a law was passed in Rome forbidding the burial of a women who died during pregnancy until the infant had been surgically removed. The two were then given separate burials. Under the Caesars, the code of laws was called Lex Caesarea, and, since the law ordered it, the operation was called the Caesarean operation. Be that as it may, it has always been regarded as a major surgical bravura which is brought to a dramatic conclusion with the presence of two individuals where previously there had been only one.

One of the earliest accepted instances of a successful outcome was an operation carried out by an Austrian sow gelder, Jacob Nufer, in 1500. The story goes that after 13 midwives had struggled in vain to deliver his wife's baby, Jacob thrust them aside and did the job himself with the aid of a few judicious slashes to the abdomen with a razor. Not only did Mrs Nufer survive, but she went on to have five more children *per via naturalis*.

About this time case studies and opinions regarding its desirability began to appear in the medical literature. The obstetrical heavies of the time opposed it on the grounds that the mother would not survive the operation. Mr Nufer was apparently not consulted. In the contemporary journals abuse of one's opponent was a common way of denigrating his worth, and the hurled insult was used as logical argument. An obstetrician called Rousset tells of a successful case of a colleague, John Lucas, who was apparently drunk at the time of surgery, and he writes, "if the operation succeeds with him when drunk, what may not he expect who perform it when sober, according to the justest rules of his art".

A specialist called Osiander said at the time that before undertaking the procedure "one should allow the patient to draw up her will and grant her time to prepare herself for death", which is hardly the frame of mind to engender when about to embark on what was supposed to be a joyous occasion. The redoubtable William Smellie took the middle view that it would be better to operate if in the perceived parlous state of the labour, mother and child are going to die anyway.

In the early days, if done at all, the uterus was not sewn up, although the abdominal wall was brought together. There were, of course, no anaesthetics then and I suppose speed of operation was important. What is more there was no thought of sterilising equipment, and intra-abdominal sutures of string or silver wire which had merely been cleaned on the lapel of the obstetrician's frock coat caused massive and often fatal infection. In fact, it was not until 1882 that the womb was sutured as a matter of course.

The first successful Caesarian operation in Britain was carried out in 1738 by midwife Mary Donally. She was an illiterate Irishwomen whose decision was untrammelled by having read the contemporary medical literature. After having had a patient 12 days in labour she felt, and rightly so, that enough was enough. So she took a razor and simply put an end to the proceedings with a few well-directed slashes. Her decision was so precipitous that she had to hold the skin edges together while someone ran to fetch needle and thread from the local tailor.

At the end of the nineteenth century Caesarean section became an elective procedure rather than one done as a last resort in heroic circumstances. As a result patients were better prepared, the operation was more leisurely and the results were quite acceptable by the lights of the day. Since then it has become a common, some say too common, and safe procedure. In fact, it is quite some time since I saw anyone running to the local tailor for thread.

One of the most important events in the history of obstetrical practice was the invention of the obstetrical forceps. It is a most remarkable story. It started in 1569 when a William Chamberlen and his wife left France to flee to England as Huguenot refugees. They brought with them their small son, Peter, and shortly after their arrival another boy was born. With singular lack of imagination, they also called him Peter. Ever after they were known as Peter the Elder and Peter the Younger.

By the early 1600s Peter the Elder prospered as a doctor due mainly to the fact that he had devised an instrument which, when placed round the baby's head in delayed delivery, could pull it out of the pelvic cavity with hitherto undreamt of ease. This apparent trick, like all good tricks, was blindingly simple; he merely made two separate blades appropriate curved to fit at once both the baby's head and the mother's pelvis. These were applied separately and then locked together when in position. Single blades had been thought of before, but not two together as it was thought they could not be opposed to act as a single unit. Chamberlen's secret was the locking device which allowed this to happen. It has not changed in principle in over 300 years.

Because it was unique, he kept the secret to himself, a state of affairs continued by other family members through four generations. Eventually this involved seven medical relatives over 125 years. It was known at the time, of course, that they had some secret weapon, and as a result over many decades several family members were in demand at some top-drawer confinements. These included being the accoucheur to Henrietta Maria, wife of Charles I, some of the mistresses of Charles II, and James II's wife. It was this last lady, you will recall, whose delivery was clouded by rumours that a male child had not appeared in the usual way, but via the back stairs, having been smuggled into the lying-in room in a warming pan.

The Chamberlen mystery itself was maintained by the simple expedient of, during delivery and supposedly in order to preserve modesty, placing a large sheet over the recumbent woman and tying the ends round the obstetrician's neck, rather like a spaghetti eater's serviette. The doctor then worked by touch, while gazing into the patient's eyes. If the instrument was needed, he reached behind into the tails of his long jacket, a sartorial trim which was fortunately fashionable at the time, where was constructed a capacious pocket in which the forceps were hidden. The unsterile instruments were withdrawn, like a rabbit out of a hat, and dexterously applied. To make sure no-one knew he was using the contentious aid, especially the rather hostile midwives ranged around, the blades were either made of wood or covered in leather so no clinking would be heard.

Both Peter the Elder and Peter the Younger used obstetrical forceps with great success. The Younger later had a son whom he also called Peter, known eventually as Peter III. This Peter in turn became a doctor, and party to the secret. In the 1650s he bought a house called Woodham Mortimer Hall near Malden in Essex. Just store that name away in your memory for it featured in the saga years later.

Peter III had a son whom in a moment of mental aberration he called Hugh. It was this man who became accoucheur to several mistresses of Charles II and Mary of Modena, the wife of that monarch's brother, James II. As Charles admitted to siring 15 illegitimate children, despite the new contraceptive invention of the royal physician, Dr Condom, Hugh must have done quite well out of coming and going into regal bedrooms.

He also wrote a book on obstetrics which refers to the fact that he and his relatives had long practised "a better way" of delivery, but carefully omitted to say what this was. But Hugh became involved in politics and had to flee to the Continent. While in Holland, and in order to raise money, he sold the family secret. Cunningly, or so he thought, he only parted with one blade of the forceps, knowing that the key of the mystery was in the way in which the two blades locked together. But the purchaser, van Roonhuysen, was a bright chap and soon worked out the rest, thus letting the cat out of the bag forever. Modifications were made over the years and forceps gradually came into general use, as indeed they are to this day.

But just cast your mind back to Woodham Mortimer Hall, property of Peter III. It was in the family from about 1650 to 1715 and then sold. Almost a hundred years later in 1813, a Mrs Kemball was visiting the place and quite by chance came across a secret hatch in an unused closet. When the door was opened inside was found a small treasure trove. There was a medallion showing the image of Charles II, a tooth wrapped in paper on which was written "my husband's last tooth", a pair of kid gloves and a box. When the box was opened there was found three pairs of the famous secret forceps in mint condition. They must have been there for 150 years or so and hidden so that no casual medical visitor would see them. They are priceless antiques now.

Some odd modifications of forceps were devised in the eighteenth century. John Burton of York constructed a set where the blades opened and closed by screwing the handle. Burton was portrayed as the grotesque Dr Slop in Laurence Sterne's *Tristram Shandy*. He brought Tristram into the world using his fiendish apparatus, thereby causing the lad to have a permanent deformity of the nose. In the tale Slop himself lacked three front teeth, which had been knocked out by his forceps when they slipped through his bloodied hands during a particularly vigorous joust he had had with the new-fangled bits of metal.

But by now the benign influence of Dr William Smellie was being felt, at least in England, and a more compassionate approach towards labouring women was taking over. The first maternity wards in the country were established in the Parish Hospital at St James's, London, in 1739 by Sir Richard Manningham. In passing, Manningham's other claim to fame was the fact that in 1726 he was sent by King George I to Godalming in Surrey to investigate the case of Mary Toft who was reputed to have given birth to 17 rabbits. He exposed the hoax, but not before Mrs Toft had become the sensation of the London season for that year, as I am sure she would have been of any other year if rumour of such a feat had got about.

The first professor of obstetrics was appointed the same year as the Rabbit-Woman sensation, and worked in Edinburgh. Mystifyingly, he had no maternity beds; they did not arrive until 1756. It was in the same city that in 1847 Sir James Young Simpson first discovered the value of chloroform as an anaesthetic in childbirth. His first heroic patient was the wife of a medical colleague. Mercifully, all went well, and the child, Wilhelmina Carstairs, was forever after called Saint Anaesthesia. Queen Victoria popularised the use of chloroform during labour when she used it during the birth of Prince Leopold, the eighth of her nine children, in 1854.

Not all obstetricians have had such felicitous results. Sir Richard Croft attended as accoucheur to Princess Charlotte. She was the only child of the Prince Regent, later George IV, and so presumptive heir to the throne. Tragically, the poor women died in labour as did the infant, whereupon Croft shot himself. That, of course, is how Victoria came to be on the throne, for she was the daughter of a younger brother of George and became heir by default.

Maternal and infant mortality have fallen dramatically in the last century, and the hazards of embarking on a pregnancy now are minuscule compared with the dangers of as recently as the pre-antibiotic days of 60 years ago. The techniques available to the accoucheur have progressed from the masterly inactivity of ancient times, through the use of jealously guarded secret instruments, to a vast array of glittering diagnostic tools available today. But the joy of a successful outcome to the obstetrician's manipulations is the same whether you were delivered by Hippocrates, Chamberlen, Smellie or the most recent graduate from medical school.

(JL)

DOCTORS AND MIDWIVES AT EACH OTHERS' THROATS

Ever since the First Fleet, midwives, or women acting as midwives, have been prominent in Australia. Phoebe Norton was transported for stealing spoons and bed linen. On the way out, she fell overboard into the Indian Ocean, was fished out, and flourished to become a busy midwife in Parramatta.

Despite their lack of status, convict women (called 'finger smiths') delivered many, perhaps most, babies. One observer wrote: "The majority of women ... were attended by women with the title midwife, but none of the associated expertise."

Of course, VIPs got VIP treatment. When Elizabeth Macquarie, wife of the Governor, went into labour in 1814, midwife Ann Reynolds was with her all afternoon. The icing on the cake was that William Redfern arrived in time for the delivery. Both of Mrs Macquarie's helpers were ex-convicts. Ann Reynolds had been transported for stealing, while the upwardly mobile Redfern was our first obstetrician, charging up to 20 guineas per delivery.

Redfern's high reputation stands in stark contrast to that of Dr John Savage, who refused to attend a woman with obstructed labour in Parramatta in 1805, allegedly because there were no instruments available. (He probably meant forceps, designed to fit around the baby's head and speed delivery.) The woman died. Savage was found guilty of neglect of duty and sent to England, where the authorities pardoned him!

Our colonial history has seen many tensions between midwives and the few available doctors (who were male).* Doctors often invoked Charles Dickens's portrait of midwife Sairey Gamp (from *Martin Chuzzlewit*):

*My account draws on the description by sociologist Evan Willis of the long-standing competition between Australian midwives, nurses, pharmacists and doctors. At times, these groups competed, not only for confinements, but also for abortions. I use the word *midwife* to mean *female midwife*.

The face of Mrs Gamp — the nose in particular — was
somewhat red and swollen and it was difficult to enjoy her
society without becoming conscious of a smell of spirits ... she
went to a lying-in or a laying-out with equal zest and relish.

While condemning midwives as untrained, ignorant and dangerous, doctors for many years would not help to train them. The law also favoured doctors. In 1864, a midwife was convicted of manslaughter; she said she had sent for a surgeon, who had refused to come because his fee was not guaranteed. The mother died, but it was the *midwife* who was blamed.

In another case near Port Fairy in Victoria, the doctor did attend as requested, but both mother and child died. On the doctor's evidence, the midwife, Charlotte Ward, was tried for manslaughter. The local paper asked "why Mrs Ward ... of much experience and irreproachable character, should be dragged before a jury and judge ... The only reasonable inference ... is that the poor woman belongs to a class which is not in favour with doctors ... " The "not guilty" verdict was very popular.

It is easy to see why people liked midwives so much. Mrs Mary Howlett attended 10–12 cases a year, and usually charged three guineas, though she was often not fully paid. She would stay and look after mother and child, as well as doing all housework for about nine days. Some midwives also used their homes as private hospitals, for confinements and sometimes for abortions. This ad appeared in the *Bathurst Times* in 1886: "Accouchements: Mrs Moir is prepared to accomodate Ladies at her residence, Pedrotta Terrace."

With blatant sexism, an editorial in the *Medical Journal of Australia (MJA)* of 1879 opposed regulation and registration of midwives, and questioned whether "the practice of obstetrics should be permitted to women at all". Yet others who were not doctors delivered babies. Thomas Sheridan, a pharmacist, delivered babies and did abortions. Within about one week, four of his aborted patients died, whereupon Sheridan went to prison for ten years. On his release, his first aborted patient died, and in 1895 Sheridan was hanged.

At least in Victoria, midwives pushing for autonomy had to fight both the doctors and the nurses. Between about 1880 and 1910, as the profession of nursing evolved, its leaders tried to include midwifery as one of its branches. Similarly, doctors wanted maternity nurses whom they could control.

Just as occurred overseas, it took decades for the findings of Ignaz Semmelweis on childbed fever to filter through into hospital practice. Devastating epidemics continued even into the twentieth century. Quite unfairly, doctors made midwives the scapegoat for childbed fever. It was the hospital doctors who could still move freely between the maternity and other sections, unwittingly spreading fever as they went, long after nurses lost this freedom.

In 1912, to boost the low birthrate, the Commonwealth Government legislated for a "baby bonus" of £5. But this generosity was very selective. It excluded not only Aboriginal and Asian women but also mothers of stillborn children, and of those who died within 12 hours.

An *MJA* editorial claimed that the lavish bonus would lead to midwives attending more births and so cause more childbed fever. The editor could not have been more wrong! Many women used the bonus to engage doctors instead: from 1913 to 1923, the proportion of births attended by midwives alone was halved. But there was little improvement in the high death rates of either mothers or babies.

One reason may have been the poor training of medical students in obstetrics and the enthusiasm of some doctors for using forceps and anaesthetics, which midwives could not use. As late as 1929, one NSW doctor boasted that, to prevent complications, he had used forceps in every one of his 768 deliveries.

In 1929, a Victorian Bill put the Nurses Registration Board in charge of midwifery. This marked (at least officially) the end of independent midwives in Victoria. Making nurses out of midwives also put them under the control of doctors.

But in the country, untrained, unregistered midwives ("rabbit-snatchers") carried on. It is reported that the main qualification of these midwives was that they had themselves delivered at least ten children who had all survived. Their cheapness was an attraction. In the Victorian town of Colac, the midwife was said to cost only two-and-a-half guineas. This compared with four guineas for the doctor, plus five guineas a week for the local private hospital.

Now in the new millennium the wheel may be coming full circle, with many doctors avoiding obstetrics, and midwives coming into their own again.

(GB)

WET NURSES

And Naomi took the child, and laid it in her bosom, and became nurse unto it.

RUTH 4:16

Each year swarms of tourists flock to the island of Capri to admire the famous Blue Grotto. Few give a thought to Tiberius, the infamous Emperor of Rome, who held unspeakable orgies there. The outraged historian Tacitus wrote: "He seized young men of ingenuous birth and forced them to yield to his brutal gratifications ... New modes of sensuality ... scandalous refinements in lascivious pleasure ..." But what made Tiberius so depraved?

Many observers blamed the alcoholic wet nurse who used to suckle him as a baby. Tiberius's successor Caligula was notorious for insanity, perversion, sadism and incest. Sure enough, his bloodthirsty wet nurse used to daub blood on her nipples before feeding baby Caligula. The legends, prejudices and controversies about breast-feeding are as old as history. We have used breast milk to treat sore eyes, blockage of urine, cataracts, burns and eczema. Direct suckling of infants by animals (such as goats, asses and ewes) is an ancient and widespread practice thought to make the baby resemble that animal. Hence goat's milk makes children swift and nimble. It was nursing by she-wolves that had made the mythical founders of Rome, Romulus and Remus, so cruel.

Kings often claimed that the gods had suckled them. In ancient Egypt, the royals themselves seldom suckled their own infants. Royal wet nurses enjoyed high status, their daughters calling themselves "milk-sister" to the king. But if the wet nurse herself delivered a daughter, so the story went, the male whom she nursed would become effeminate.

In the second century, the Greek physician Soranus set out the ideal qualities of a wet nurse. (These ideals changed little until the nineteenth century.) She must be cheerful, active and healthy. Her breasts and even the blue veins on them must be not too large or too small. Her teeth must be strong and white, since bad teeth and bad breath would affect the baby's lungs. She should speak well and not stammer. The teacher Sir Thomas Elyot (1544) even wanted nurses who could speak good Latin to the infant.

She should have had (and hence be immune to) smallpox and measles. But she must not have gout, leprosy, falling-sickness (epilepsy), consumption, squint or bladder stones. Also taboo were red hair, freckles, vaginal discharge, drinking or smoking, and above all menstruation and pregnancy. Many accounts forbade a menstruating woman to wet nurse, because they believed that periods meant poor, scanty milk. The story went that during pregnancy and lactation, all the blood normally lost during menstruation went via a duct to the breasts, where it changed into milk. Indeed, old anatomy texts used to show a duct leading from the uterus to the breasts. Many also insisted that a nurse who became pregnant must stop suckling at once.

Between about 1500 and 1800, many medical and religious writers gave wet nurses a bad press. Walter Harris (1689) wrote: "The passing bell hardly ever ceases ringing out the death of infants which have died for the neglect, nastiness, barbarity, or intemperance of their nurses." Sir John Floyer (1706) said bluntly: "No child has the rickets unless he has a dirty slut for his nurse."

Why make scapegoats of wet nurses? Before about 1800, people believed that the wet nurse (like the breastfeeding mother) could transmit to the baby her own diet, ideas, beliefs, intellect, speech and all other physical and emotional qualities. What affected the nurse affected the child. Anything bad that happened to the child was the nurse's fault. Nor was it only the ignorant who held such beliefs. Throughout his long life, Samuel Johnson blamed his tuberculosis and poor eyesight on his wet nurse.

So what did you do for a sick nurse or a sick baby? First give the nurse a purge, then look at her diet. If all else failed, sack the nurse and find another, or else wean the baby. But occasionally a nurse got another chance. In 1727, when the son of the Earl of Cardigan was six months old, his nurse fell and broke her forearm. "I sent immediately for Dr Foyer who came and set it. We do not suffer Master Robert to suck for these five or six days, for these things are always attended with a fever." After a few days: "The nurse is entirely free from pain and has not been at all feverish — so that in a day or two the child may suck without any manner of danger."

In the seventeenth and eighteenth centuries, a major cause of death among infants was "overlaying", where a seemingly healthy baby was found dead in the morning. People blamed these deaths on nurses falling asleep on top of the babies and so suffocating them. Overlaying occurred not just in Britain but wherever wet nurses were common. In Florence, wet nurses had to use a wooden frame — the *arcutio* — to protect babies. Wrote Valerie Fildes: "Every nurse ... is obliged to lay the child in it, under pain of excommunication. The *arcutio*, with the child in it, may be safely laid entirely under the bed cloaths in the winter, without the danger of smothering."

But the British knew better. Though the *arcutio* was widely advertised in European journals, it never caught on in Britain. Modern research suggests that unless the nurse or mother is drunk, drugged or unconscious, even without a protective frame, it is very hard to overlay or suffocate a healthy baby. So did some of these babies die of what we now call SIDS or cot death?

(GB)

Chapter 14

A FINAL WORD: LOUIS PASTEUR — THE GREATEST

If you look back on the panorama of the history of medicine stretching over six or seven thousand years, there has emerged a pantheon of men and women who stand out above the others; their contributions have been outstanding and far-reaching. They have appeared in all eras and, although in many cases their worth was not recognised until years later, they set the tone. I suppose many have been rather like medical Van Goghs, appreciated when safely out of the way. For this distinguished role a very few are regarded by common consent to be ahead of the other greats, the crème de la crème. To attain this enviable position the laureates have had to make a contribution that has significantly changed the course of medical history, or indeed history itself.

Who are these few? Rather as when picking an all-time great Olympic team, opinions vary as to the worth of individual excellence, so my choice for the award of bronze and silver medals in these lofty stakes could be reasonably questioned. But my gold goes to someone who has few detractors.

I would award the bronze to Claudius Galen (c.130–c.210). I give it not because he was correct in his thinking (on the contrary he was often wrong) but because his influence was such that it lasted almost unquestioned for almost 1500 years. In casting about for answers he laboured under the disadvantage that he was not permitted to dissect humans, only animals, so often his conclusions were inaccurate in detail. Despite this he was an acute observer, voluminous writer and powerful advocate of his ideas. His influence on medical thinking was without parallel in history, both in its diversity and certainly in the length of time it held sway.

My silver would go to William Harvey (1578–1657). As we have seen, by dissection and careful use of the so-called "scientific method" of proving all his observations by repeating and cross-checking their accuracy, he exploded the old theories of Galen. By so doing he opened up for the first time ever the logical process of medical thought as we know it today. His most famous contribution, of course, was the discovery of the circulation of the blood in the mid-1600s. After him came the deluge.

But the gold belongs, in my opinion, to the man who discovered the basic cause of those infections which had from the beginning of recorded time decimated the population of the world — Louis Pasteur. Pasteur (1822–1895) was neither medically qualified nor trained — he was a chemist — but for my part I believe he was the greatest benefactor to medical welfare in human history.

Pasteur was born in the village of Dole in the Jura district of eastern France on 27 December 1822. His father was a retired sergeant major from Napoleon's *grande armèe* and at the time of the birth was working as a tanner. At school Louis was regarded as an average student but a very good drawer. He went to the Royal College of Franche Comte in Besancon and graduated as a Bachelor of Letters (Arts) in 1840 and as a Bachelor of Science in 1842.

From there he went to Paris, where for the first time his academic ability began to be recognised. He wrote theses in chemistry and physics, but his real interest lay in crystallography, and in 1848 he read a paper in this subject to the Academy of Science. Shortly afterwards he became Professor of Physics at Dijon, and in 1849 he was elevated to the Chair of Chemistry at Strasbourg where he was able to pursue his interest in crystals. At the time he was 26 years old.

In 1853 Pasteur succeeded in transforming tartaric acid into racemic acid, a piece of fundamental chemistry which showed for the first time that the biological properties of substances depend not only on the nature of the atoms, but also on the manner they are arranged in space. For this work he was awarded the red ribbon and rosette of a Knight of the Legion of Honour.

The following year he was made Professor and Dean of the Faculty of Science at Lille where he became a sought-after lecturer who was able to display clarity of thought yet lose none of the polished and well-modulated delivery which made him a favourite among the students. He also introduced a new teaching aid whereby he summarised his notes and had them bound together to give to the students. He also introduced evening classes for the first time ever so that undergraduates could attend after doing a day's work elsewhere. About 300 students were attracted to his sessions.

It was during these lectures at Lille that his famous aphorism first saw the light of day — "In the field of observation, chance only favours the mind which is prepared." Or as it is usually put, "Opportunity presents to the prepared mind."

In 1862, while still at Lille, he was elected a member of the Academy of Science, mainly as a result of his fundamental work on the fermentation of alcohol, that arcane path of winemaking never far from a Frenchman's heart. He felt that diseases of wine might be caused by "microscopic vegetations which would develop in certain circumstances of temperature ... the alterations of wine are coexistent with the multiplication of microscopic vegetations". These conclusions became known as the "Pasteur effect" and were to stand him in good stead later in his medical experiments. At that time they were used to overcome the wine problem by the use of heat, or "pasteurisation" as the locals had it. It was applied to milk as well and the word, of course, has passed into common usage.

As can be imagined, some old diehards thought the vintage would be ruined by heat. Ageing, it was claimed, required dark, cool cellaring and time. Pasteur replied, "The ageing of wines is due, not to the fermentation, but to a slow oxidisation which is favoured by heat."

Rather foolhardily he elected to test his theory on the navy. Two barrels of wine were sent to sea, one with wine which had been previously heated, and one unheated. Perhaps rather surprisingly, after ten months at sea, both barrels arrived back intact. The heated one was found to be "limpid and mellow" and the other "limpid but astringent". The discovery was implemented commercially and the export of French wines increased enormously.

The scientist then applied himself to the question of how these micro-organisms which caused fermentation arose. Was it spontaneous generation, the contemporary popular favourite explanation, or were they present in the air all the time, just waiting to settle? By means of simple experiments involving filtered air and exposure of unfermented liquids to pure air high in the Alps, Pasteur proved that food decomposes when placed in contact with germs in the atmosphere and does not spontaneously generate new organisms within itself.

Proving the existence of these "germs" in the air led Joseph Lister (1827–1912), a surgeon working in Scotland, to apply the same principle to surgically infected wounds. The results were dramatic. This single discovery of attempting to eliminate the germs transformed surgical operations with their terrible risks and appalling side effects due to infection into procedures with an acceptable degree of danger; not good, but distinctly better. Thus much misery was relieved and countless lives were saved.

If Pasteur had thrown light on no other problem during his life, this single event itself would have put him in the running for the gold. But there was much more to come. At the time the lucrative business of silkworm cultivation, centred in Alais in southern

France, was threatened by a disease which was killing off the worms. Pasteur was called in, and again with the aid of his microscope and keen observation detected the cause to be in the micro-organisms in the gut of the silkworm. He saved the industry and the honour of France.

By now, in 1868, the scientist was at the top of his form, or so everyone thought. He was in great demand and became the Professor of Chemistry at the Sorbonne University in Paris. Then disaster struck. On 19 October that year and when he was 45, he felt a strange tingling sensation on the left-hand side of his body but insisted on reading a paper he had promised to give to the Academy of Science. That night the threatening cerebral haemorrhage occurred and the master became paralysed down the left side. Fear was expressed for his life, and 16 leeches were prudently applied behind the ear to relieve the pressure. Obviously, some areas of medicine had not received his searching attention. It's a pity he did not have some surgical problem so some of his new-fangled ideas could be tried out. In any event, the attack was regarded as being so serious that the French Emperor daily sent a footman to enquire after his progress.

Fortunately, the great mind remained lucid, but officialdom was not so sanguine as regards the ultimate outcome, for work was ordered to be stopped on the laboratory which the government was currently building specifically for him. It now seemed that the trouble and expense might turn out to be unnecessary.

Slowly and painfully, Pasteur recovered and at length he returned to the silk-growing region to continue his investigations. He finished them with, as I say, complete success.

In the following years he looked at the fermentation of yeast within the brewing industry, both in France and in London. His work led him to modify the brewing of beer so that it would not deteriorate with time and hence could travel to distant countries for sale.

In 1877 he investigated so-called "splenic fever" in sheep; there was no stopping the man. This is known now as anthrax and at the time was threatening to wipe out the pastoral industry as cattle and horses were also affected. Modifying the guidelines of Edward Jenner (1749–1823) with his treatment of smallpox initiated a hundred years before, Pasteur prepared a dilute form of the causative germ which he then injected into the animal thereby causing a mild attack and thence immunity. As we have seen, his method was actually more in line with the ideas of Lady Mary Wortley Montagu rather than Jenner. Chicken cholera was approached in the same way with equally gratifying results. It almost seemed he could do no wrong and the accolades kept coming.

By now he was more a bacteriologist than chemist, and indeed in 1873 was elected to the Academy of Medicine despite his lack of formal medical training. As a product of this honour, Pasteur started to look at the diseases of man, and the hospital began to assume as great an importance in his life as did the laboratory. So it was at this comparitive late stage in his professional life, at the age of 51, that he approached for the first time those problems for which he is best remembered.

He looked at pus under the microscope and identified bacteria, recognising its key role in infection. One day at the Academy when the discussion turned to puerperal fever, the deadly infection of recently delivered women, his colleagues were ventilating their various theories when Louis Pasteur suddenly rose and said, "None of those things cause the epidemic; it is the nursing and medical staff who carry the microbes from an infected woman to a healthy one". If that was so, someone remarked, it would never be found. Whereupon Pasteur went to the board, drew a diagram of a chain-like organism and said, "There, that's what it is like". Nobody had seen such a thing before. Nonetheless, their collective professional feathers were ruffled after being told by a man who had never even been into a lying-in ward that they should be sterilising the linen and washing their hands between cases.

Others had no such doubts about his genius, however, and honours were heaped upon him by a grateful government and indebted industry. He represented his country at international conferences; he had dinner with the Prince of Wales; a commemorative plate was placed on the house where he was born; the daughter of the famous missionary Dr David Livingstone sought him out to present a book on her father's life, and in 1882 he was accorded what was regarded as his country's greatest honour, membership of the Académie Française.

From all over the world came appeals and requests for consultations. Many thought him a physician with almost mystical powers, but as a scientist of the time said, "He does not cure individuals; he only tries to cure humanity". Pasteur himself regarded his laboratory as "the temple of the future".

Despite all this hullabaloo and advancing years, the scientist's most famous work was still in front of him. It was to be a piece of research which in the end Pasteur himself placed above all others, for it was concerned with a mystery which had constantly haunted him for many years — that of hydrophobia, as he called it, or rabies.

It was known that the frightening condition was caught from affected dogs, and in 1880 the researcher was presented with two such "mad dogs" by an Army veterinary surgeon who was trying to find a cure for the puzzle of hydrophobia. Incredibly, the contemporary method of prevention was to file down the teeth of the dog so its bite would not penetrate the skin. The only clinical certainties known then were that the rabies virus was present in the saliva of "mad" animals, that it was passed on by bites and that incubation could be from a few days to several months. The animals were regarded as "mad" on account of the bizarre behaviour which overtook them when infected.

The symptoms of the condition were restlessness, muscle spasms, a fierce thirst accompanied by an inability to swallow (hence "hydrophobia", fear of water), convulsions and fits of rage. The victims often died by drowning in their own saliva. The whole presented as a very distressing picture.

Pasteur set to work on dogs. By observing the troubled behaviour in an infected canine he felt the virus must be lodged in the brain, and by microscopic examination set out to prove this.

It did not take a researcher of this man's quality long to find that his supposition was right. He then cut up a dog brain and cultured the organism within the specimens in rabbits, concentrating it through several generations. Brain tissue was taken from these infected smaller animals and dried to attenuate the germ. The now-inactive tissue was crushed, mixed with water and injected under the skin of a healthy dog. Attenuated viruses of more and more virulent strains were injected over several days. The animals were then exposed to the bite of an infected dog, or they were injected with the deadly virus, while the experimenter anxiously waited to see what happened.

In fact nothing happened; the disease was resisted. Another day, another triumph.

The thought following on from this was to inoculate all the dogs of France to give them immunity. As there were an estimated 2 500 000 such animals, Gallic enthusiasm took on a new meaning. Indeed, the only thing which seemed to dampen this zeal was not the enormity of the task but the non-availability of sufficient rabbits to provide the vaccine emulsion.

But the real problem lay with the disease in humans. Celsus, the first century Roman physician, writing at about the time of Christ, described the symptoms, saying, "The patient is tortured at the same time by thirst and by an invincible revulsion towards water." He went on to recommend cauterisation of the bite with a red hot iron and the application of various corrosives. In passing it should be noted that this fear of water is characteristic of the infected human and is not replicated in the dog.

Pliny the Elder (23–79 AD) recommended eating the livers of mad dogs as a cure, and Galen, for reasons best known to himself, suggested crayfish eyes. Such remarks put his bronze medal in jeopardy. In the eighteenth century sea-bathing held sway, especially at Dieppe. In some less squeamish quarters the ultimate solution of suffocating the victim, presumably as a perceived act of kindness, was advanced as a treatment. In the time of Pasteur's childhood the turn of cautery had come round again, and hot needles were even thrust into the face while nitric acid was used elsewhere. No wonder the sight of a salivating, disturbed dog held great terror in a population where the treatment was patently worse than the complaint.   .

So after 2000 years treatment had come full circle. Then Louis Pasteur appeared and all was soon to be light. By mid-1885, when he was 62, the experimenter was ready to try his theories on the human state, and on 6 July that year opportunity literally knocked at the door in the form of nine-year-old Joseph Meister and his mother. They were from Meissengott in the Alsace region where the boy had been bitten two days previously by a rabid dog. The animal in question had promptly been shot by the owner and when the stomach was opened hay, straw and pieces of wood were found. When the lad's parents heard this they feared the worst and consulted their doctor. He cauterised the wounds with carbolic and suggested that Joseph should be taken to Paris to see Pasteur, which the parents promptly did.

As he opened the door and took a look at the frightened boy, Pasteur knew he had reached his rubicon. The scientist was touched, for although he had seen people seeking his advice many times before, here was a forlorn victim with 14 wounds and in so much pain he could hardly walk. Could he risk a form of treatment which thus far had only been tried on dogs? Doubtless, he sucked his teeth and absentmindedly examined his cuticles as he pondered the dilemma. He sent the pair away so he could agonise without the emotional pressure of being fixed with the appealing gaze of a sick child.

Two colleagues were consulted. What was there to lose, they asked. If, they continued, the certain danger which threatened the lad was weighed against the chances of snatching him from death, then the experimenter should see it was more than a right, it was a duty — high-sounding and facile sentiments if you do not have to take responsibility.

Another look at the child decided him. It was now or never. A 14-day-old vaccine was chosen, ensuring that no living virus was within. It was injected into the site of one of the bites. The prick was slight and young Joseph did not move.

A few days later on 11 July the boy was sleeping and eating well. The virulent strength of the continuing daily inoculations increased, and Pasteur suffered from mounting anxiety, knowing there was no drawing back now. He stopped all other work and had difficulty sleeping, and when he did drop off had nightmares of children dying from suffocation.

The serum grew stronger over a ten-day period. On the last day the strength was of a degree that would have killed a rabbit in experimental circumstances. But as Louis Pasteur tossed all night, Joseph Meister slept the sleep of the innocent.

The treatment was completed and mother and son went home, while the therapist went to a deserted country house with his daughter in order to settle his jangled nerves. The feared telegram from the local GP never came, and 31 days after the bite, Joseph was declared out of danger.

Pasteur returned to Paris to resume his life of experimentation, doubtless going about his task with a warm feeling of a job well done. With justification, he was now regarded as "The Oracle" and a steady stream of visitors beat a path to his door. Bottles of wine were brought to study and "cure" of acidity. Silkworms, cases of swine fever and anthrax and chicken cholera all showed up.

Inevitably, another case of human rabies presented. Although lacking the uniqueness of being first, the fearlessness and audacity of this second victim, together with the success of the subsequent treatment, so caught the public imagination it is commemorated to this day by a statue set in the grounds of the Pasteur Institute in Paris.

The story is this. On 24 October 1885 the mayor of Villers-Farley in the Jura district wrote to Pasteur with the information that a shepherd boy had been bitten the previous day by a mad dog. It seems six young shepherds were guarding their sheep when a large dog approached. Five ran for safety, but the eldest, 14-year-old Jupille, turned to protect his fellows and wrestle with the animal. It seized his left hand but he managed to disengage himself by forcing the jaws open with the right hand. Consequently, both hands were severely bitten and covered in the deadly saliva. However, the dog was eventually overpowered and clubbed to death. A necropsy was done the next day and rabies was confirmed. It was at this point that the mayor wrote to Paris.

Pasteur replied immediately to say he had treated only one human case, and, although that one had been successful, the dangers were great. The family decided to give it a go, but in the inevitable and agonising delay, six whole days had elapsed before Jupille arrived in the laboratory. With young Meister it had been two-and-a-half days, which had been stressful enough, but following the spectacular success of that case, the master had a newfound confidence and plunged straight into treatment.

A few days later he went to speak at the Academy of Science and recounted the Meister case, his original brief for speaking. But while on his feet Pasteur also seized the opportunity to tell of the current problem. He spoke of the courage of the young shepherd protecting his companions and almost sacrificing his life in the process. Other speakers took up the theme and the upshot was that for his courage and devotion 14-year-old Jupille received the Montyon prize of France's highest national institution, the Académie Française.

The boy recovered and later, as I say, a statue was erected to the shepherd in the grounds of the Pasteur Institute. It shows the boy fighting off the rabid dog and is still there for all to see and marvel at.

For Pasteur's part he was lionised wherever he went. Prizes and praise continued to be heaped upon him. The President of the Academy of Science said of the scientist's efforts with rabies, "it is one of the greatest steps ever accomplished in the medical order of things ... treatment for a disease the incurable nature of which was a legacy handed down from one century to another. It is to M. Pasteur that we owe this, and we could not feel too much admiration or too much gratitude for the efforts on his part which have lead to such a magnificent result." The speech contains much more of this kind of hyperbole.

When news of Pasteur's paper regarding the two cases became common knowledge he was besieged by victims from all sides. He prepared vaccine from the spinal marrow of rabid dogs. The specimen hung in a flask to be gradually dried out by the action of caustic potash lying in the bottom of the glass. The marrow was then cut up and placed on a glass plate, and veal broth was added and the mixture was ground up. The vaccinal fluid was now ready. It was left for 14 days, and each patient had a series of little glasses with different strengths.

The hour for inoculation was 11 a.m. and Pasteur himself was always present. It was the master himself who called the patients in. One child, Louise Pelletier, was brought along 37 days after being bitten. Pasteur hesitated to treat her in case the almost certain failure after so long an incubation period would put off other suffers. In the end, common humanity prevailed, and he sat at the girl's bedside as, despite all his efforts, she began to fit. Louise died and when Pasteur came downstairs he burst into tears. Her death was the only one in his first 350 cases. The contemporary untreated death rate was 16 per 100.

Four children came from New York to be treated, having been financed by public subscription via the good offices of *The New York Herald*. They all returned in triumph. Master Jupille kept up a correspondence with Pasteur and he wrote back with political correctness encouraging the young shepherd always to take advice from his father and mother and not waste time with other boys!

The Academy of Science established a unit for the prevention and treatment of hydrophobia. This was the previously mentioned Pasteur Institute, and it was built in 1888 in what had been a market garden in the middle of Paris. Where young lettuces and spring onions had grown there arose a solid stone building with a suitably opulent Louis XIII façade, grand enough to go with its irreproachable international reputation. It is still there and functioning as a research laboratory but with a greatly enlarged

The "Action-Storyteller" Pasteur at work at the
Academy of Science, recounting how the 14-year-old
escaped the attack from the rabid dog. He was also
known to spin a damn fine yarn over a glass of milk as well.

brief, its current prime research being into AIDS. It can be visited, along with the reconstruction of Pasteur's living quarters and his tomb in the basement. It was to this institute that the Duchess of Windsor deemed that the proceeds of the posthumous sale of her jewellery should go.

Pasteur had a few detractors who thought that his failures were not being published. These upset him and his health began to suffer. Despite being cheered by resounding support from English workers, he had two minor strokes in 1887, and although he recovered, his pale face and gaunt frame testified to the stresses of the previous few years. But he lived on, a legend in his own lifetime.

On 27 December 1892 a distinguished academic gathering was held at the Sorbonne University in Paris to celebrate the maestro's

70th birthday. The place was packed with the most illustrious men and women of the era, both from France and from overseas. A medal was struck showing, on one side, Pasteur in profile wearing his skull cap, and on the other an inscription which read, "To Pasteur on his seventieth birthday. France and Humanity grateful."

Louis Pasteur mounted the podium leaning on the arm of the President of the Republic of France, no less. The speeches were grand, sonorous and numerous. "Who can now say how much human life owes to you and how much it will owe to you in the future." "The day will come when another Lucretius will sing, in a new poem of nature, of the immortal Master whose genius engendered such benefits." "Pasteur's works shine with such dazzling light, that, in looking, one is inclined to think that it eclipses all others." And so on, and so on, and so on.

Pasteur was too weak and emotional to speak, so his reply was read by his son. It was a fairly short presentation by the standards of the occasion, but during it there occurred one magical instant which not only made a great impact then but has remained to be remembered as one of science's most deathless moments.

Among those present was Joseph Lister, Lord Lister, who, as we have seen, years before had applied the germ theories of Pasteur in his surgical practice and had thereby saved countless lives. As Pasteur spoke of the overseas visitors he said, "You bring me the deepest joy that can be felt by a man whose invincible belief is that Science and Peace will triumph over Ignorance and War ... that the future will belong to those who will have done most for suffering humanity. I refer to you, my dear Lister..." Spontaneous applause and shouts of "Vive Pasteur!" broke out, and as the crowd rose cheering, Lister and Pasteur embraced. This electrifying moment marked the sealing of the new era in medicine and surgery.

Louis Pasteur lived for another three years, becoming gradually weaker but remaining mentally very active. Work in his laboratory was carried on by his assistants, many of whom became famous in their own right. It was from them that antitoxins for diphtheria and plague were discovered. Indeed, the tomb of Pasteur's successor, Dr Emile Roux, inventor of the diptheria serum injection, is in the garden of the institute, and the road on which the building stands is now called Rue du Docteur Roux. Pasteur took a critical interest in all their efforts, but the fire had gone and on 28 September 1895 he died.

We may smirk at the grand and fulsome praise that was heaped upon him, but there is no doubt of his greatness. The fundamental discoveries of Louis Pasteur transformed modern medicine, surgery, obstetrics and public health and made possible the control of infectious disease. He brought about a revolution in scientific investigation by studying the agents of disease in their natural environment and almost always provided a solution. He had insatiable curiosity, an analytical and mercurial mind, a remarkable gift of observation, and immense and persistent enthusiasm. While gracious in his dealings with his colleagues, he was not immodest and did not refuse the honours heaped upon him, and for good reason. For, of course, he knew what we now know — he was simply the greatest.

(JL)

BIBLIOGRAPHY

Chapter 1: Pioneers

Appleby, L., *A Medical Tour Through the Whole Island of Great Britain*, Faber & Faber, London, 1994.

Birch, C. A., *Names We Remember*, Ravenswood, Beckenham, 1979.

Gethyn-Jones, J. E., *The Jenner Museum*, Jenner Trust, Bristol, 1986.

Gordon, R., *The Alarming History of Medicine*, Sinclair Stevenson, London, 1993.

Guthrie, D., *A History of Medicine*, Thomas Nelson & Sons, Edinburgh, 1945.

Haggard, H. W., *Devils, Drugs and Doctors*, Blue Ribbon Books, New York, 1929.

Karlen, A., *Plague's Progress*, Indigo, London, 1995.

Keynes, G., *The Personality of William Harvey: The Linacre Lecture*, Cambridge University Press, Cambridge, 1949.

Leavesley, J. H., *Medical Byways*, ABC/Collins, Sydney, 1984.

Perry, C. B., *Edward Jenner*, Jenner Trust, Bristol.

Porter, R., *The Greatest Benefit to Mankind*, HarperCollins, London, 1997.

Roberts, S., "Constance Stone — Australia's first woman doctor", *Journal of Medical Biography*, vol. 3, 1995, pp. 1–7.

Chapter 2: Frauds

Broad, W. & Wade, N., *Betrayers of the Truth*, Simon & Schuster, New York, 1982.

Hixson, J., *The Patchwork Mouse*, Doubleday, New York, 1976.

Youngson, R. & Schott, I., *Medical Blunders*, Robinson, London, 1996.

Chapter 3: Versatile doctors

Encyclopaedia Britannica: Micropeadia, "Buccaneers", vol. II, 1975.

Encyclopaedia Britannica: Macropaedia, "David Livingstone", vol. 11, 1975.

Gelfand, M., *Livingstone the Doctor: His Life and Times*, Basil Blackwood, Oxford, 1957.

Longfield-Jones, G. M., "Buccaneering doctors", *Medical History*, vol. 36, 1992, pp. 187–206.

Ma, K., "Sun Yat-sen (1866–1925)", *Journal of Medical Biography*, vol. 4, 1996, pp. 161–170.

Manne, R., *The Petrov Affair: Politics and Espionage*, Pergammon Press, Sydney, 1987.

Osler, W., *Michael Servetus*, Oxford University Press, Oxford, 1909.

Roland, C., Introduction to S. Squire Sprigge, *The Life and Times of Thomas Wakley*, Krieger, Huntington, NY, 1974 (1899).

Sheridan, R., "The doctor and the buccaneer", *Journal of History of Medicine and Allied Sciences*, vol. 41, 1986, pp. 76–87.

Sun Yat-sen, *Kidnapped in London*, Arrowsmith, Bristol, 1897.

Chapter 4: Medicine and the arts

Beighton, P. & Beighton, G., *The Man Behind the Syndrome*, Springer-Verlag, London, 1986.

Boswell, J., *Boswell's London Journal 1762–1763*, William Heinemann, London, 1950.

Boswell, J., *Life of Johnson*, Oxford University Press, Oxford, 1980.

Burgess, A., *Ernest Hemingway and His World*, Thames & Hudson, London, 1978.

Gibbon, E., *The Decline and Fall of the Roman Empire*, Penguin, London, 1981.

Gordon, R., *The Literary Companion to Medicine*, Sinclair Stevenson, London, 1993.

Haggard, H. W., *Devils, Drugs and Doctors*, Blue Ribbon Books, New York, 1929.

Keynes, M., "The personality and illnesses of Wolfgang Amadeus Mozart", *Journal of Medical Biography*, vol. 2, 1994, pp. 217–32.

Landon, H. C. R., *1791: Mozart's Last Year*, Flamingo, London, 1989.

Leavesley, J. H., *Medical Byways*, ABC/Collins, Sydney, 1984.

MacLaurin, C., "Gibbon's hydrocoele", *Medical Journal of Australia*, 20 April 1920, pp. 385–7.

McLeish, K. & McLeish, V., *Composers and Their World: Mozart*, Heinemann, London, 1978.

Meyers, J., *Hemingway: A Biography*, Macmillan, London, 1985.

Morton, R. S., *Gonorrhoea*, W. B. Saunders, London, 1977.

Murray, T. J., "Dr Samuel Johnson's movement disorders", *British Medical Journal*, vol. 1, 1979, pp. 1610–14.

Page, P. M., *Medical Biographies*, University of Oklahoma Press, Norman, 1952.

Porritt, A. E., "Some historical surgical operations", *General Practitioner of Australia and New Zealand*, 16 December 1937, pp. 288–94.

Porter, R., "The dark side of Samuel Johnson", *History Today*, vol. 34, December 1984, pp. 43–6.

Porter, R., *The Greatest Benefit to Mankind*, HarperCollins, London, 1997.

Power, D., "Some bygone operations in surgery: A historical lithotomy: Samuel Pepys", *British Journal of Surgery*, vol. XVIII, no. 72, April 1931, pp. 541–5.

Schultz, M. G., "The strange case of Robert Louis Stevenson", *Journal of the American Medical Association*, vol. 216, no. 1, 1971, pp. 90–4.

Vincent, E. H., "And so to bed", *Surgery, Obstetrics and Gynaecology*, vol. 87, no. 3, September 1948, pp. 353–7.

Chapter 5: Opium

Inglis, B., *The Opium War*, Hodder & Stoughton, London, 1976.

Chapter 6: Famous and infamous people

Brander, M., *The Victorian Gentleman*, Gordon Cremonesi, London, 1975.

Fuentes, C., *The Buried Mirror: Reflections on Spain and the New World*, Andre Deutsch, London, 1992.

Green, V., *The Madness of Kings: Personal Trauma and the Fate of Nations*, Alan Sutton, Phoenix Mill, England, 1993.

Innes, H., *The Conquistadors*, Fontana/Collins, London, 1969.

King, G., *The Murder of Rasputin*, Century, London, 1966.

Leavesley, J. H., *A Mixed Medical Bag*, ABC/Collins, Sydney, 1985.

L'Etang, H., *The Pathology of Leadership*, Heinemann, London, 1969.

Magnus, P., *King Edward the Seventh*, John Murray, London, 1964.

Maples, W., *Dead Men Do Tell Tales: The Strange Cases of a Forensic Anthropologist*, Doubleday, New York, 1994.

Palmer, A. (ed.), *Age of Optimism*, Weidenfeld & Nicolson, London, 1970.

Winton, R., *From the Sidelines of Medicine*, Australasian Medical Publishing Company, Sydney, 1988.

Chapter 7: Death

Benitez, M. B., "A 39-year-old man with mental changes", *Maryland Medical Journal*, vol. 45, 1996, pp. 765–9.

Cavendish, R. "Death of Edgar Allen Poe", *History Today*, vol. 49, no. 10, October 1999, p. 52.

Daugherty, C. G., "The death of Socrates and the toxicology of hemlock", *Journal of Medical Biography*, vol. 3, 1995, pp. 178–82.

Day, D., *John Curtin: A Life*, HarperCollins, Sydney, 1999.

Fabricant N., *13 Famous Patients*, Pyramid, New York, 1960.

Hamilton. E. & Cairns. H. (eds), *Plato: The Collected Dialogues*, Princeton University Press, Princeton, 1961.

L'Etang., H., *Ailing Leaders in Power 1914–1994*, Royal Society of Medicine Press, London, 1995.

MacGregor Burns, J., *Roosevelt, Soldier of Freedom 1940–45*, Weidenfeld and Nicolson, London, 1971.

Ross, L., *John Curtin: A Biography*, Melbourne University Press, Melbourne, 1977.

Serle, G., *For Australia and Labor: Prime Minister John Curtin*, Prime Ministerial Library, Perth, 1998.

Wilkins, R., *The Fireside Book of Death*, Robert Hale, London, 1990.

Chapter 8: Diseases

Fraser, D. W. et al., "Legionnaire's Disease", *New England Journal of Medicine*, vol. 297, no. 22, 1977, pp. 1189–96.

Hudson, R. P., "Lessons from Legionnaire's Disease", *Annals of Internal Medicine*, vol. 90, 1979, pp. 704–7.

Steinschneider, A., "Prolonged apnoea and sudden death syndrome: Clinical and laboratory observations", *Pediatrics*, vol. 50, 1972, pp. 646–54.

Thomas, B. & Gandevia, B., "R. Francis Workman and the history of taking the cure—consumption in the Australian colonies", *Medical Journal of Australia*, 4 July 1959, vol. II, pp. 1–10.

Chapter 9: Disasters and eccentrics

Alexander, L., "The commitment and suicide of King Ludwig II of Bavaria", *American Journal of Psychiatry*, vol. III, 1954, pp. 100–7.

Bateson, C., *The Convict Ships, 1787–1868*, 2nd edn, Brown, Son & Ferguson, Glasgow, 1969.

Beattie, O. & Greiger, J., *Frozen in Time: The Fate of the Franklin Expedition*, Bloomsbury, London, 1987.

Blunt, W., *The Dream King*, Hamish Hamilton, London, 1970.

Cobcroft, M., "Medical aspects of the Second Fleet", in *Australia's Quest for Colonial Health*, eds J. Pearn & C. O'Carrigan, Department of Child Health, Royal Childrens Hospital, Brisbane, 1983.

Cookman, S., *Ice Blink*, Wiley, London, 2000.

Dods, L., "Early paediatrics in the Antipodes", *Medical Journal of Australia*, 10 June 1961.

Gregg, C., *A Virus of Love and Other Tales of Medical Detection*, Scribner's, New York, 1983.

Hay, G. G., "The illness of Ludwig II of Bavaria", *Psychological Medicine*, vol. 7, 1977, pp. 189–96.

Hughes, R., *The Fatal Shore*, Pan, London, 1987.

Maltby, J. R. & Lee, J. A., "The medical establishment and association with unqualified practitioners: The sad case of Doctor Axham", *Journal of Medical Biography*, vol. 3, 1955, pp. 119–23.

Newman, E., "The strange case of King Ludwig II of Bavaria", *The Saturday Book*, No. 4, Hutchinson, London, 1944.

Nicol, B., *Behind the Myth*, ABC Enterprises, Sydney, 1989.

Pearn, J. (ed.), *Milestones of Australian Medicine*, Amphion Press, Brisbane, 1994.

Roe, Francis, "Medicine and books", *British Medical Journal*, 9 June 1979, pp. 1553–4.

Sexton, C., *The Seeds of Time: The Life of Sir Macfarlane Burnet*, Oxford University Press, Melbourne, 1991.

Sauerbruch, F., *A Surgeon's Life*, translated by Fernand G. Renier and Anne Cliff, Deutsch, London, 1953.

Sunday Times, *Suffer the Children: The Story of Thalidomide*, Andre Deutsch, London, 1979.

Thorwald, J., *The Dismissal: The Last Days of Ferdinand Sauerbruch, Surgeon*, Thames & Hudson, London, 1961.

Youngson, R. & Schott, I., *Medical Blunders*, Robinson, London, 1996.

Chapter 10: Quirks and oddities

Klawans, H. L., *Trials of an Expert Witness: Tales of Clinical Neurology and the Law*, Bodley Head, London, 1991.

Klawans, H., *The Medicine of History*, Raven, New York, 1982.

Lyons, A. & Petrucelli, J., *Medicine: An Illustrated History*, Macmillan, Melbourne, 1979.

Strean, H. S., *Behind the Couch: Revelations of a Psychoanalyst*, Wiley, New York, 1988.

Chapter 11: Discoveries

Pflaum, R., *Grand Obsession: Madame Curie and Her World*, Doubleday, New York, 1989.

Watts, G., *Pleasing the Patient*, Faber & Faber, London, 1992.

Chapter 12: Mayhem and murder

Wambaugh, J., *The Blooding*, Bantam, London, 1989.

Sakula, A. "Roderigo Lopez: Executed 1594", *Journal of Medical Biography*, vol. 3, 1952, pp. 114–18.

Chapter 13: Midwives, women and babies

Fildes, V., *Breasts, Bottles and Babies*, Edinburgh University Press, Edinburgh, 1986.

Guthrie, D., *A History of Medicine*, Thomas Nelson & Sons, London, 1945.

Haggard, H. W., *Devils, Drugs and Doctors*, Blue Ribbon Books, New York, 1929.

Jeffcoate, T. N. A., *Principles of Gynaecology*, Butterworth, London, 1957.

Willis, E., *Medical Dominance: The Division of Labour in Australian Health Care*, Allen & Unwin, Sydney, 1989.

Chapter 14: A final word: Louis Pasteur — the greatest

Guthrie, D., *A History of Medicine*, Thomas Nelson & Sons, London, 1945.

Porter, R., *The Greatest Benefit to Mankind*, HarperCollins, London, 1997.

Vallery-Radot, R., *The Life of Pasteur*, vols I & II, Archibald Constable, London, 1902.